LOCOMOTION PAPERS

THE
PEAK FOREST TRAMWAY

including
THE PEAK FOREST CANAL

by
David Ripley

THE OAKWOOD PRESS

© David Ripley, 2007
First Edition 1968
Second Edition 1973
Third Enlarged Edition 1989
Fourth Revised Edition 2021

ISBN 978-0-85361-749-5

Printed by
Blissetts Unit E1-E8 Shield Drive, West Cross Ind Pk, Brentford, TW8 9EX

References

Farey Survey of Derbyshire, 3 vols 1815
Bradshaw Inland Navigation, 1903
Portus, C. *Peakland*
Rhodes, E. *Peak Scenery*, 1824
Rees, A. *Cyclopaedia*, 1804
Cox, Rev. *Derbyshire*, 1896
Hutchinson *Tour of the Peak*, 1822
Bunting, W.B. *Chapel-en-le-Frith*, 1940
Dow, G. *Great Central Railway*, Vol. 1, 1959 (subsequently reprinted by Ian Allan Ltd)
Carter, E.F. *The Railway Encyclopaedia*, 1967
Rimmer, A. *The Cromford & High Peak Railway*, 1956
White, F. *History of Derbyshire*, 1857
Baxter, B. *Stone Blocks and Iron Rails*, 1966
Hill, E.L. *The Peak Forest Canal*, 1967
Bellhouse, M.A.L. *The Story of Combs, My Village*, 1968
Denby Marshall, C.F. *A History of British Railways down to the year 1830*. Pub. 1938
Lee, C.E. *The Evolution of Railways*, 1943

Articles in the following Journals and Newspapers.
Derbyshire Countryside, Edgar Allen News, Railway Magazine, Buxton Advertiser, The High Peak News, The Derby Mercury, The High Peak Reporter, The Derbyshire Times.

The author also acknowledges the assistance of amongst many:
Jack Buxton, Herbert Green, Sam Eyre, Phillip Murray and W. Cartledge (all of whom have now passed away), Ron Cooke, and Ruth M. Brace for the photographs from her survey of the line. Ferodo Limited Chapel-en-le Frith, ICI Limited, Mond Division; the staff of The Derbyshire County Library Service.

Publisher's Note: David Ripley submitted revisions for the fourth edition of the Peak Forest Tramway to Oakwood Press in 2005. These were mainly to chapter six to update the changes since 1989. By 2005 publishing had changed and had become digital, while the book was on sheets of film, which made updating and reprinting it difficult. The book has now been digitized and redesigned so that his revision can be included, some small additions have been made to update it to 2021. In the intervening years, David suffered a stroke, and entered care; unfortunately he died in 2020 before this book could be reprinted.

Published by
The Oakwood Press, 54-58 Mill Square, Catrine, KA5 6RD
Telephone: 01290 551122   Website: www.stenlake.co.uk

Contents

	Forwords to Second and Third Editions	4
One	Early History	7
Two	The Track and Rolling Stock	11
Three	Traffic and Operation	19
Four	The Route Described	23
Five	The Peak Forest Canal	57
Six	Conclusion	67

Appendices

One	Chronology of the Peak Forest Tramway and Canal	71
Two	Some employees of the Canal, Tramway and Private Haulage Teams	77
Three	Lime Kilns and Lime Burners	78
Four	Other Lines that served the Canal	79

The Peak Forest Canal showing the junction of the main line and Whaley Bridge branch. The horses used an underpass to get from one side of the "canal main line" to the other.
Oakwood collection

Foreword to the Second Edition

Since the first edition of this booklet was published, several interesting additional items have come to light, partly due to the interest caused by the booklet, and partly by the researches of Brian Lamb and the late Phillip Murray.

Some of this additional material has been included in the following pages.

Mention must also be made of the considerable works that are being carried out on the Peak Forest Canal by two Voluntary Societies, firstly the Inland Waterways Protection Society, who are restoring the Buxworth Basin. At the time of writing 600 yards of what was once derelict canal has been restored to pristine condition; not restored to the Cruising Waterway Standard as set out in the 1968 Transport Act, but to its original depth of 5 ft 6 in., this being 1 foot more than was usual on 18th century canals, as the top section was to act as an additional water storage.

As a result of their efforts, the Inland Waterways Protection Society was in 1970 awarded a Bronze Plaque under the European Conservation Year, Countryside Awards Scheme.

Work on cleaning out the Basin proper has now commenced and already the remains of one of the Tramway Wagons has been unearthed; no doubt much more of the canal and its tramway will be revealed.

The other Society working on the Canal is the Peak Forest Canal Society; this society is restoring the flight of 16 locks at Marple, and with luck and a little help they hope to have the locks in working order within the next two years.

If you are in the area, spend a little time at Buxworth and Marple, look at the works that are being restored, and with a little imagination you should be able to recapture the heyday of the canal building age.

D. Ripley
Chapel-en-le-Frith,
October 1972.

Foreword to the Third Edition

It is more than 20 years since the first edition of The Peak Forest Tramway was published and during this period the area has seen many changes.

The lower Peak Forest Canal has been re-opened from its junction with the Ashton Canal to its meeting with the Upper Peak Forest and Macclesfield Canals at Marple. Mention must be made of the splendid works carried out by the Peak Forest Canal Society during the 1970s in restoring the flight of 16 locks at Marple, and the mammoth task of cleaning out the canal to enable cruising to take place once more on the Cheshire Circuit.

At Bugsworth, the Inland Waterway Protection Society has worked continuously since 1968 on cleaning and restoring the Basin, walls have been rebuilt, a replacement bridge over the first arm has been constructed, cobbles relaid and thousands of tons of rubbish removed. Help in recent years by a Manpower Services Team has speeded the efforts of the many volunteers who have given so freely of their time and labour.

In 1970 the Inland Waterways Protection Society was awarded a Bronze Plaque under the European Conservation Year, Countryside Awards Scheme.

Then in 1980 a by-pass for Chapel and Whaley Bridge received Government approval, the route of which was to follow very closely that of the Peak Forest Tramway from Chapel to Bugsworth. A local enquiry was held and the Inspector confirmed that the route should follow the Blackbrook Valley just as Outram had done for the Tramway some 150 years previously. The bypass is now open, the valley has been changed, some of the items associated with the canal and its tramway have gone, but enough remains for anyone with a little imagination to recapture the age of canals and their associated tramways.

D. Ripley
Chapel-en-le-Frith
December 1988

Note: The name of Bugsworth changed to Buxworth around 1932.

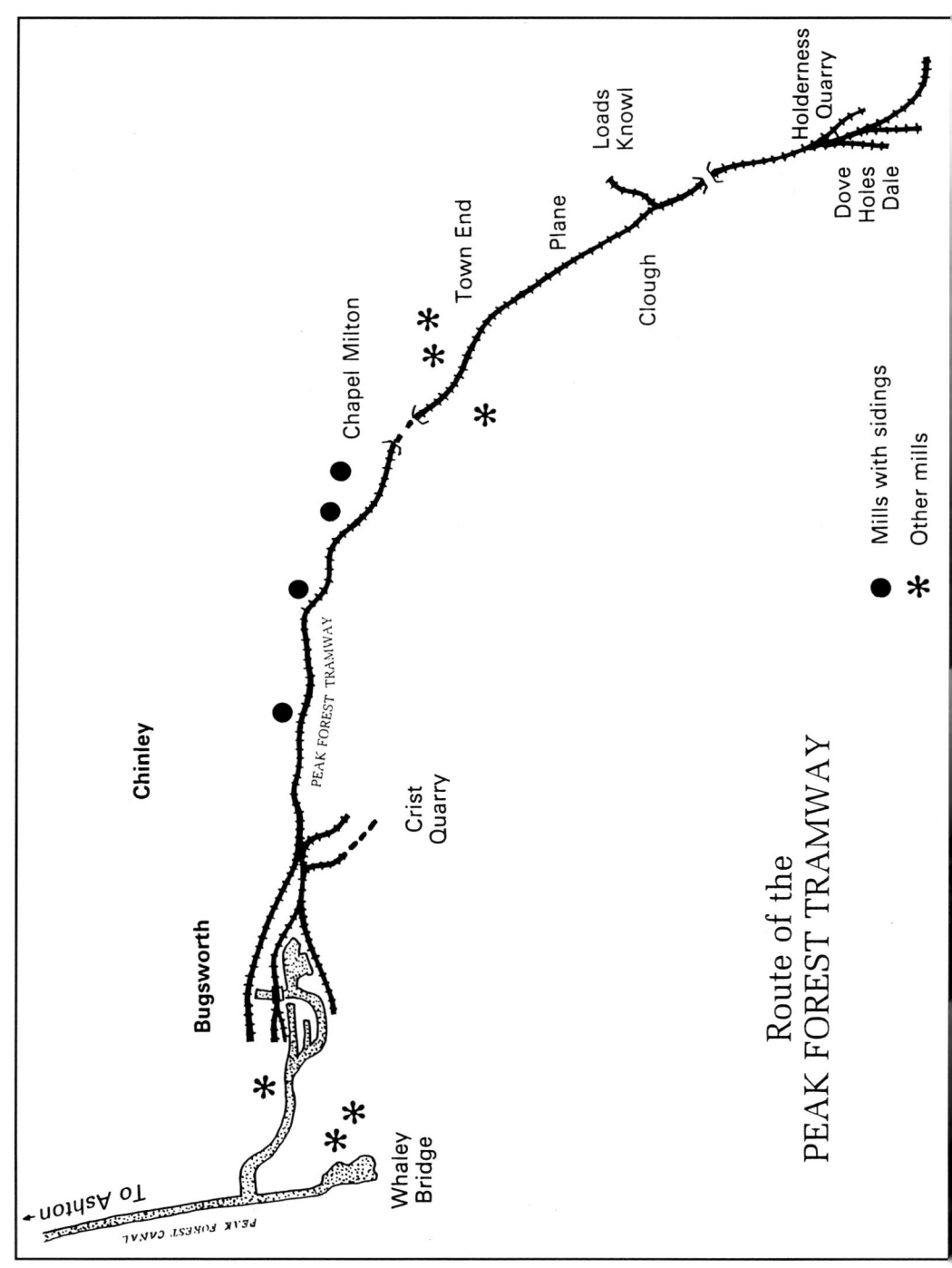

Chapter One

Early History

Following the success of the Bridgewater Canal, built by the Duke of Bridgewater with James Brindley, (born at Tunstead now part of Peak Forest Parish) as his engineer, business men on both sides of the Pennines looked to canals to solve one of their major problems, that of transporting their wares to markets. Many schemes were put forward during the second half of the 18th century, a period during which Brindley was involved in construction as diverse as the Trent and Mersey Canal, with its many locks and a long tunnel at Harecastle where it crosses the watershed between the North and Irish Seas, the Chesterfield Canal, designed to provide an outlet for the coal and iron trades of North East Derbyshire, and the contour-hugging Oxford Canal.

In 1786 a group of business men representing the Manchester and Sheffield regions met to discuss a link between the towns, several schemes were proposed, but due to the hilly terrain of the Peak District none of the schemes came to fruition. By 1790 Manchester was a "hub" for canals and a navigable waterway, with the original Bridgewater Canal from Worsley, the new Bridgewater to Preston Brook where it joined the Trent and Mersey Canal and also made contact with the River Weaver by the locks at Runcorn, and the old Mersey and Irwell River Navigation; this 'hub' was to receive many more spokes during the next two decades.

The next year saw a scheme for the proposed Ashton and Rochdale Canals and in 1792 a rough draft and survey was prepared for a canal to link Cromford to Manchester. This scheme for a canal to follow the valleys of the Derwent and Wye rivers came to nought, due to opposition from the wealthy landowners in the Bakewell and Buxton areas. Several of the backers for this scheme now changed their objectives: the chemical industries which were expanding along the River Weaver and the Sankey Canal were short of one important raw material, limestone, and limestone was quarried in the Peak District.

The preliminary survey suggested a canal to Chapel Milton with a plateway from there to the start of the limestone plateau at Loads Knowle in the Parish of Peak Forest, the canal to have two flights of locks, one flight at Marple and the second between Bugsworth and Chapel Milton. Reservoirs to provide water for the locks would need to be constructed at Chapel-en-le-Frith and in The Wash.

> Plans for the canal, tramway and reservoirs were prepared by T. Brown and deposited with the Clerks of the Peace for Derbyshire and Cheshire and the Bill

presented to Parliament. It was approved the following year and on 28th March, 1794, the Peak Forest Canal Act allowed the Peak Forest Canal Company to be incorporated.

The preamble to the Act gave the reasons for the canal and tramway as: ... making and maintaining a navigable Canal from out of the Canal Navigation from Manchester to or near Ashton under Lyne and Oldham in the County Palatine of Lancaster, at the intended Aqueduct Bridge in Duckinfield in the County of Chester to or near to Chapel Milton, in the County of Derby and a communication by railways or stone roads from thence to Loads Knowle, within Peak Forest in the said County of Derby and a branch from out of the said intended canal to Whaley Bridge in the said County of Cheshire.

The company was to be known as the "Company of Proprietors of the Peak Forest Canal" and was authorized to raise £90,000 in shares of £100 each and up to £60,000 on mortgage of the rates. Although T. Brown had carried out the survey, the company appointed Benjamin Outram as Engineer for both the canal and tramway with T. Brown as Residential Engineer.

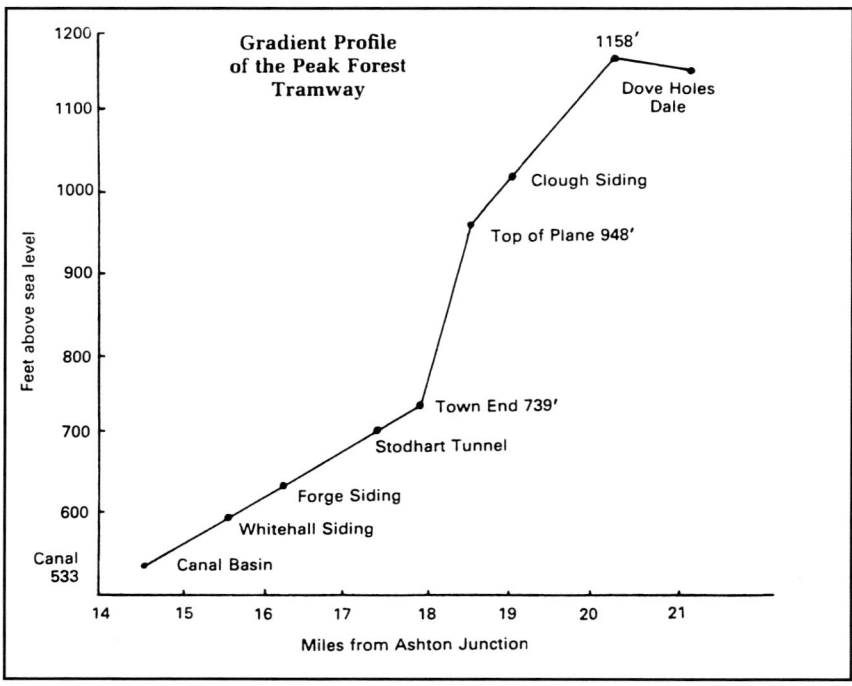

Benjamin Outram was born in the south of the County near Alfreton in 1764, his place of birth having been described as a substantial stone grey house in King Street, his father James being an iron-master and surveyor. He was named Benjamin after Benjamin Franklin who at that time was looking after the interests of the State of Pennsylvania in London. Franklin was a friend of the family and acted as his godfather.

It is recorded that at the Ripley foundry owned by James in 1775 were cast 'L' section plates, and these were laid by his 'plate-layers' at the Duke of Norfolk's Colliery the following year.

Benjamin grew up in an engineering environment and in 1789 he was working with the experienced William Jessop on the Cromford Canal, by the side of which Outram purchased a small estate belonging to Buttley Hall. A considerable amount of Outram's work was connected with tramways and he is regarded as a pioneer in the field. He constructed several in his native County: one was the "Little Eaton" running from the Little Eaton arm of the Derby Canal to the Drury Low colliery at Denby, which opened in 1795 and was five miles in length. By the end of the century it was carrying upwards of 40,000 tons of coal and stone a month and showed an annual profit of £1750. Another Outram line was built to connect the quarries and lime works near Calke Abbey to the Ashby Canal. This, the Ticknall line was 4 miles in length, and cost £31,600 to build, but the returns from the line were very good, and it continued to show a profit until the 1890s.

Other major works of his included the survey and construction of the Standedge tunnel on the Huddersfield Narrow Canal, at 5,415 yds in length the longest tunnel on an artificial waterway in Britain. His services were also in demand in South Wales; a pamphlet exists, entitled "Report and Estimate of the Proposed Rail-ways from the Collieries in the Forest of Dean to the Rivers Severn and Wye" by Benj. Outram, Engineer. This was published in 1801 and the following year he was associated with the Sirhowy Tramway Company and the Monmouth Canal.

Such was his influence on the area that T.G. Cunning writing in 1824 tells us that,

> ... as late as the year 1790, there was scarcely a single railway in all South Wales, whilst in the year 1812 the rail-ways in a finished state, connected with canals, collieries, iron and copper works in the Counties of Monmouth, Glamorgan and Carmarthen alone, extended to upwards of 150 miles in length, all of these railways were constructed on the tram-plate principle, using stone blocks instead of wooden sleepers.

With William Jessop, Francis Beresford and John Wright, the latter a member of the Wright banking family who had married Francis Beresford's daughter, he established the firm of Outram & Co., later the Butterley Company. An ironworks was established which produced much of the track used on the various tramroads and railways with which the proprietors were associated.

Outram married in June 1800, much to the surprise of his friends. His bride, Margaret Anderson a beauty and a capable woman was the daughter of Dr. James Anderson, the Scottish scientist, agricultural pioneer and writer. She bore him two sons, the second of whom joined the army and as General Sir James Outram, became known as the "Bayard of India".*

Benjamin died tragically whilst in London on business in 1805 at the early age of 41, after a very active career as an engineer and surveyor. His was a short life, but most of the canals and tramways he was connected with were still working a century after his death, and several canals are still used today.

The Butterley Company prospered and by the 1920s it owned in addition to the ironworks, forges and rolling mills in Condor Park, railway wagon works, brickworks, limestone quarries and nine collieries with a potential output of over 4,500,000 tons of coal per annum.

---

* Sepoy Mutiny of 1857. Generals Havelock and Outram with about 1000 men relieved the force defending the British Residency in Lucknow on 25th September, 1857 after it had been under siege since 1st July. The force was not however strong enough to break out with the injured and infirm. so the combined force of less than 2000 men held out against the rebel army of 60,000. General Havelock became ill and General Outram took sole charge and such was his inspiration that his small force held out until the main British army under Sir Colin Campbell finally broke through to Lucknow on 17th November, 1857.

Chapter Two

The Track and Rolling Stock

The tramway was laid according to Outram to a gauge of 4 ft 2 in. between the flanges, this, it is believed, is similar to the gauge used elsewhere by both Benjamin Outram and his father being influenced by the types of farm and mining carts already in use around Ripley. Whilst his partner William Jessop favoured "edge" rails similar to those still in use today, Benjamin favoured the "L" section rail which his father had cast for several years.

The first rails used on the Peak Forest were 3 ft in length with a tread 3 inches wide and ¾ inch thick, with a hump backed flange. They were supplied by James Outram and a letter of 3rd June, 1797 to Browne of Dissley exists asking for payment of rails supplied for the Peak Forest Tramway. Originally the rails were laid directly on to the stone blocks and had a countersunk notch or slot at their ends; two lengths, when placed together on the stone blocks, were held by an iron spike with a flat point and a rectangular head shaped to fit into the notches on the

Town End Wharf in 1903. View towards the tunnel under the Buxton Road. There were extensive sidings and stores at the foot of the incline, seen in the right hand top corner of the photograph. *Courtesy I. Green*

A section of stone sleeper blocks on the derelict course of the Peak Forest Tramroad.
Courtesy T.P.Keeley

Standard mineral wagon on a section of track. (Note the stone sleeper blocks and slippers holding the rails.)
Oakwood collection

THE TRACK AND ROLLING STOCK 13

Section of track still in situ at Bugsworth. *Author's collection*

Diagrammatic sketch of the "Main Line" switch and track sections.

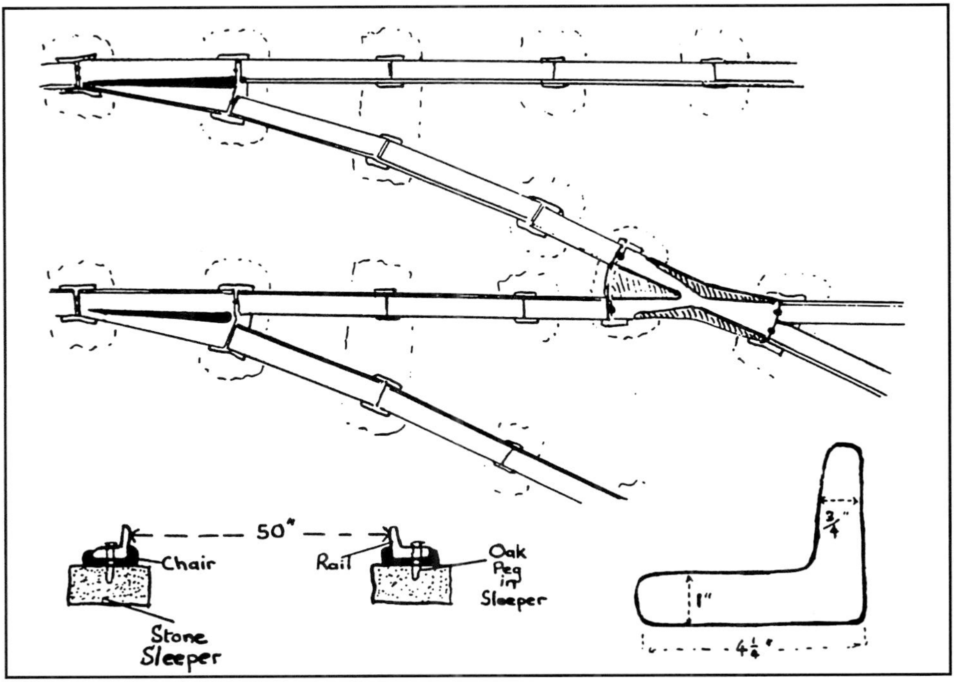

rails. This was then driven into an oak plug which had been placed in the hole in the sleeper. Outram specified octagonal oak plugs some 5 inches in length, and for the rails to be cast thicker at their ends to enable them to seat more firmly.

By the 1830s rail breakages made the canal company consider relaying the track and in 1833 a new rail with a high flange was ordered. These were not made to fit directly on to the stone sleepers but on to a slipper, saddle or pedestal made of cast-iron. The rail was now spiked through the saddle. A meeting in the Minute Book in 1833 states that all plates and pedestals were to be examined at Bugsworth before they were sent up the line. In the following year some 4,010 yds is recorded as having been relaid with new rails and blocks and a further 3,158 yds repaired.

Rail breakages remained a problem and in the 1860s the track was again relaid, this time using steel rails of the same section and weight as the 1833 cast-iron rails, produced at the Gorton Works of the Manchester, Sheffield and Lincolnshire Railway Company who by this time owned the canal and tramway. This time, round spikes appear to have been used.

At road crossings Outram had specified that level crossing rails were to be of double thickness. A flat "U" section approximately 3¾ inches between the flanges and weighing 90 lb. to the yard was used. Whilst work was in progress on the by-pass in 1985, a temporary roadway was constructed at Bugsworth and on clearing the site a length of the "U" section track was uncovered. The rails were sunk into the roadway so as to cause as little obstruction as possible to road vehicles.

The saddles were of various types; some have been found with three hole fixings (involving three holes being drilled in the stone blocks) and one line of thought for these is that they were used where the track was on a tight bend. This is partly borne out by the location of stone blocks with three holes found some years ago on the Chapel Milton to Stodhart section. A standard saddle was approximately 8½ in. by 7½ in. wide.

Rails were fastened together after 1835 by means of fishplates and the saddles were so cast that their flanges did not catch the fish plates, some later saddles seem to have been fabricated from wrought-iron. Packing plates were used under rails and saddles where necessary to ensure that the track was level.

Pointwork was constructed from numerous components which all had to be cast; several different lengths of rail, switch rails, crossover rails and saddles have all been found in recent years or are in museums. Unfortunately the section preserved in the Railway Museum at York is not a point that was taken from one location, instead it is a composite of several points.

The wagons, or as they would be termed in the 1790s "waggons", were of very simple construction similar to that used for farm carts: wooden frames, wooden axles, with iron fittings to take flangeless wooden wheels hooped with iron bands and the wheels held on to axles by means of a large washer and a flat lynch pin. Bodies were constructed so that they could be lifted off the wagon chassis complete with load. The main type in use was constructed of sheet iron, low sided with three fixed sides and one end closed by a "gate". Other types in use included water carts, these being fashioned from timber and caulked in a similar fashion to the narrow boats on the canal. Flats were used for the carriage of provisions and farming supplies.

An improved type of chassis, in use by the 1820s, employed cast-iron wheels, iron stub axles, and wrought-iron fittings; this type of wagon weighed between 26 and 30 cwt. and could carry a load of 2 to 3 tons. The main use of this type was the carriage of lime and limestone, the wheel base of the wagons varying from 2 ft 9 in. to 3 ft 6 in. At either end of the chassis large hasps were fitted, these were used with stout iron

Remains of old high-sided wagon. *Author's collection*

chains to join the wagons into trains and also on the wharf side in conjunction with the "tippling wheels" to unload the wagon.

As the line had several curves, friction caused a build-up of heat in the wheels which would seize on the axle or become brittle and break. One solution adopted by the men who ran the trains of wagons was to fit "drip cans", placed one over each wheel and consisting of a simple tin can with a thin neck filled with water. The open neck was plugged with cotton waste so that the water could slowly percolate on to the wheel to reduce rail friction. The wheels were of cast-iron weighing about ¾ cwt. each, running loose on the axles.

The bed of the tramway between Bridgeholm and Chapel Milton in the 1950s. The stone sleepers can be clearly seen. This part of the tramway is now used as a test track for brake testing by Ferodo Ltd. The viaducts carry the former Midland line from Manchester to Derby, now used solely for mineral traffic from the Dove Holes and Buxton area.

Ferodo Ltd.

THE TRACK AND ROLLING STOCK 17

Standard mineral wagon. *Ferodo Ltd*

Detail of wagon wheel and braking bar. *Ferodo Ltd*

In the late 1940s all that remained of the Top Basin was the main stump of the crane (a standard type as used throughout on the Peak Forest Canal). *J. Morten*

Chapter Three

Traffic and Operation

The reason stated in the preamble to the Peak Forest Canal Act for the construction of a canal and tramway, was the conveyance of lime and limestone from the Peak Forest area to industry throughout Lancashire and Cheshire. We know that for over 120 years limestone travelled down the tramway to Bugsworth where it was transshipped to journey on by the canal network. As the advantages of the canal and its tramway were realized by the owners of mills and forges in the area, steam power began to replace water power to drive looms etc., coal being imported from originally the Cheshire and Lancashire pits and later, when the Macclesfield Canal was open, from those in Staffordshire.

By the 1880s three boats arrived every two days with coal from various collieries near Poynton and this was then carried up the line to feed the lime kilns in Dove Holes Dale.

Farey in 1816 gives us an insight into the trades that had developed near the Canal and Tramway: tan yards at Chapel, woollen mills at Whaley Bridge, basket and whisket (a basket made from straw) making at Chapel, a bar slitting mill at Bugsworth, calico weaving at Chapel, and cotton spinning and weaving at Brownside, Bugsworth, Chapel (2 mills), Dove Holes, New Mills (7 mills), Peak Forest and Whaley Bridge, at which latter place mills also produced tapes.

Heavy industry, in the form of hammer mills and forges where bar iron was made from "Pigs", is listed for Bugsworth, Chapel-en-le-Frith and Chapel Milton.

It is doubtful if much of this industry would have prospered without the economical transport provided by the tramway and canal system. Many of the mills that were started during the first decade of the 19th century are still in use some 170 years on, albeit not for their original purposes.

When the canal opened, the canal company operated the tramway with its own staff and horses and this system carried on into the 20th century, augmented where necessary by "Private Teams". The teams of horses provided by the company were stabled to serve the various sections of the line, and it was claimed that men and horses were never moved from one section to another, but served their working lives in one place. For the top section, horses were originally stabled at Laneside Farm, Dove Holes and when they needed to be re-shod, the blacksmith at the "Top of the Plane" would make and fit new shoes. In the 1840s the horses were transferred to new stables that had been built further down the Dale near the "New Line" Quarries.

The Blacksmith's Shop at the "top of the plane" c.1905; the building still exists today.
Author's collection

A "nipper" with his horse outside the Blacksmith's Shop around the turn of the 20th century.
Author's collection

On the bottom section, horses were stabled at Town End, Chapel and Bugsworth, shoeing being done by the blacksmiths situated at Town End and at Crist Quarry.

As the bulk of the traffic was from the quarries and lime kilns in Dove Holes to the canal at Bugsworth, the route being on a falling gradient, it follows that gravity played a large part in the working of the tramway. Both two- and five-horse teams were used, the five-horse teams hauling the empty trains of wagons up the line and half trains loaded with coal for the lime kilns. The two-horse teams mainly dealt with the sundries traffic, corn, salt, provisions and agricultural implements etc. In 1860 the canal company owned 35 horses, whilst a further 17 belonged to private hauliers.

A team had a "waggoner" in charge and he was assisted by a "nipper", a nipper being a boy of 12 or 13 when he started with the team. Their duties included both gravity-descending trains and use of the horses to the best advantage to take empties and loads back up the line. In 1830 it is recorded that a wagoner was paid 12s. per week, this had increased to 20s. by the 1880s. In 1903 the Great Central Railway Company raised the wagoner's wage to 25s. and the nipper received 15s. When the nipper reached the age of 17 he was transferred to other work. As masons and joiners received wages comparable to that of the wagoners, it can be assumed that the wagoner was considered to be "skilled".

On the sections above and below the inclined plane, the loaded wagons were formed into trains of from 16 to 40 wagons and run down by gravity. The method of controlling and braking the descending trains was unique to the line, and hair-raising to say the least: in today's world, the Railway Inspectorate would have closed it down.

Whilst the nipper led the horses down, making sure that both he and his charges kept clear of the gravity trains on the down line, the wagoner rode down with the train. He would have to set the train in motion, and then quickly "sprag" as many wheels as he considered necessary to control the train's speed, before he boarded it by standing on the lynch pins and holding on to the side of the wagon. The "sprags" were sometimes called "locks" and consisted of a short chain with a hook at each end, one hook being hung on the wagon side and the other, (which was somewhat flattened) being made to engage in the spokes of the wheel thus preventing it from turning.

If the train was travelling too fast, the wagoner would endeavour to brake it by hooking a spare lock on to the side of the wagon and throwing the flattened hook in to the wheel, causing it to skid and thus act as a brake. In addition to spare locks, the wagoner also carried one or more "sticks", some 5 feet in length and normally made of hickory. These were used to lever a derailed wagon back on to the line, or, as frequently happened, to change a broken wheel.

Broken wheels were the main cause of breakdowns and spare wheels were placed every few yards alongside the track. When a broken wheel occurred, the wagoner used his sticks to lever the wagon into an upright position, packed the axle level with stones, then removed the lynch pin and old wheel. A new wheel was located and fitted, the lynch pin replaced, the stones removed from under the axle, and with a push, the wagon was away once more.

The use of "locks" is mentioned briefly by George Taylor when writing in the *Great Central Railway Journal* in December 1902 and has been confirmed to the Author by three gentlemen who had at one time been employed on the tramway. They point out that if the stick had been used to brake the trains as several people have suggested in the past, the action of pushing it through the wheel would have knocked the wagoner off as he rode on the lynch pins. Furthermore it would have fouled the traffic on the adjacent line, or have caught the horses and nippers legs as they walked on the line side. Some times a wagoner would use his stick as a brake to keep a train stationary in a siding; then· when he pulled it out the train would start to move without much effort.

The *Buxton Advertiser* for 5th November, 1888 has a report on an inquest into a quarry man who was thrown off a wagon and killed. It seems that he was "conducting a wagon down an incline when he improvized for a brake in the shape of an iron bar, the wagon jumped and the deceased was thrown to the ground".

At the incline the trains were broken up and lowered down the incline in batches of eight or so wagons (approximately 30 tons gross) attached to one end of an endless cable. The descending wagons drew up either eight (or more) empty wagons or a lesser number of those carrying coal etc., to Dove Holes, the weight of the load down in all cases exceeding that travelling up.

As the number of wagons being hauled "up" varied it was necessary to have some method of controlling the speed and also to stop the wagons when they reached the bottom of the incline. This was done by a brakesman in an elevated "tower" at the top of the incline who could see the whole length of the incline and the signal at the foot of the plane. Full details of the Plane and its working are given in the next chapter.

Chapter Four

The Route Described

Today you can no longer walk the length of the tramway, parts have been lost to road improvement schemes and some for industrial use, but if we could turn the clock back to, say, the years following the First World War, we could have experienced a pleasant and at the same time instructive walk. Let us imagine that we are taking that walk.

Starting from the canal basin at Bugsworth, with its many sidings and buildings, the line goes east alongside the Blackbrook, connecting lines from Crist Quarries join on the right hand and soon we reach Whitehall Works. This works has been a paper mill, a cotton weaving mill and a bleach works.

Leaving the works which had its own siding we cross the public roads twice in the hamlet of Whitehough, on past the "Forge" again with its own siding, alongside the stream, cross Charley Lane and past Bridgeholm. The line now swings right under the viaducts, one of which once carried the main Midland Railway line from London to Manchester

A view of Bugsworth, with limestone awaiting shipment on the wharf (to the right of the bridge) whilst the frame to carry the rails over the wharf to the "tippler" is clearly visible on the left. *Author's collection*

Looking at the Top Basin (from the bridge) with the "tippler" wheel on the left and the transfer warehouse on the right.   *Author's collection*

Stone crusher house at Bugsworth. It was on the south side of the North Arm and the building collapsed when the original crusher was removed to a small quarry in Eccles.
Author's collection

via Derby. We are now in Chapel Milton which was to have been the terminus of the canal if the original survey had been built. Then we turn right and follow the Chapel-en-le-Frith to Glossop Turnpike Road for almost a hundred yards, before it swings left and enters the tunnel which takes it under the road and the garden of Stodhart House until it re-appears, and crosses the little Blackbrook on a fine stone bridge. We have now reached Chapel-en-le-Frith, the largest place on the route and having a population of 2,667 when the tramway was opened.

Passing several tan pits and into a large marshalling yard which stretches on both sides of the Chapel to Castleton Turnpike Road, a short tunnel under the Buxton Road follows and we are at the foot of the steep incline, known locally as the Plane. After a steep climb to the top we reach a further marshalling yard with a brakesman's hut straddled across the lines like a sentry on duty.

The route now turns east once more and we begin the gentle ascent to Barmour Clough, the road from Chapel-en-le-Frith to Buxton at first well below us but climbing sharply to reach us by the Clough corner. Now the line swings right and to our left we can see the Loads Knowl, the original end of the line. Running side by side for a few yards with the Buxton extension of the Stockport, Disley and Whaley Bridge Railway we reach Dove Holes, then passing under the road once more we reach Dove Holes Dale and its many quarries and lime kilns.

Having very briefly described the route of the tramway I will outline in greater detail its various component parts which can be visited and examined today.

Bugsworth Basin

"Bugsworth is a thriving inland canal port", is how a guidebook of the last century starts its description of Bugsworth and it continues, "The Basin at Bugsworth holds upwards of twenty long boats; it has a loading shed and stables for upwards of forty barge horses".

For a township of less than three hundred people, Bugsworth was a busy and comparatively affluent area in the early part of the nineteenth century. The canal had brought trade, its industries, mills, quarries, coal mines and lime kilns were all developed during this period giving the local people a choice of work; many had small farms and did part-time work for the canal company. All the goods which arrived by canal had to be registered at the canal office where the boat was weighed. The method of weighing employed involved the boat passing through a

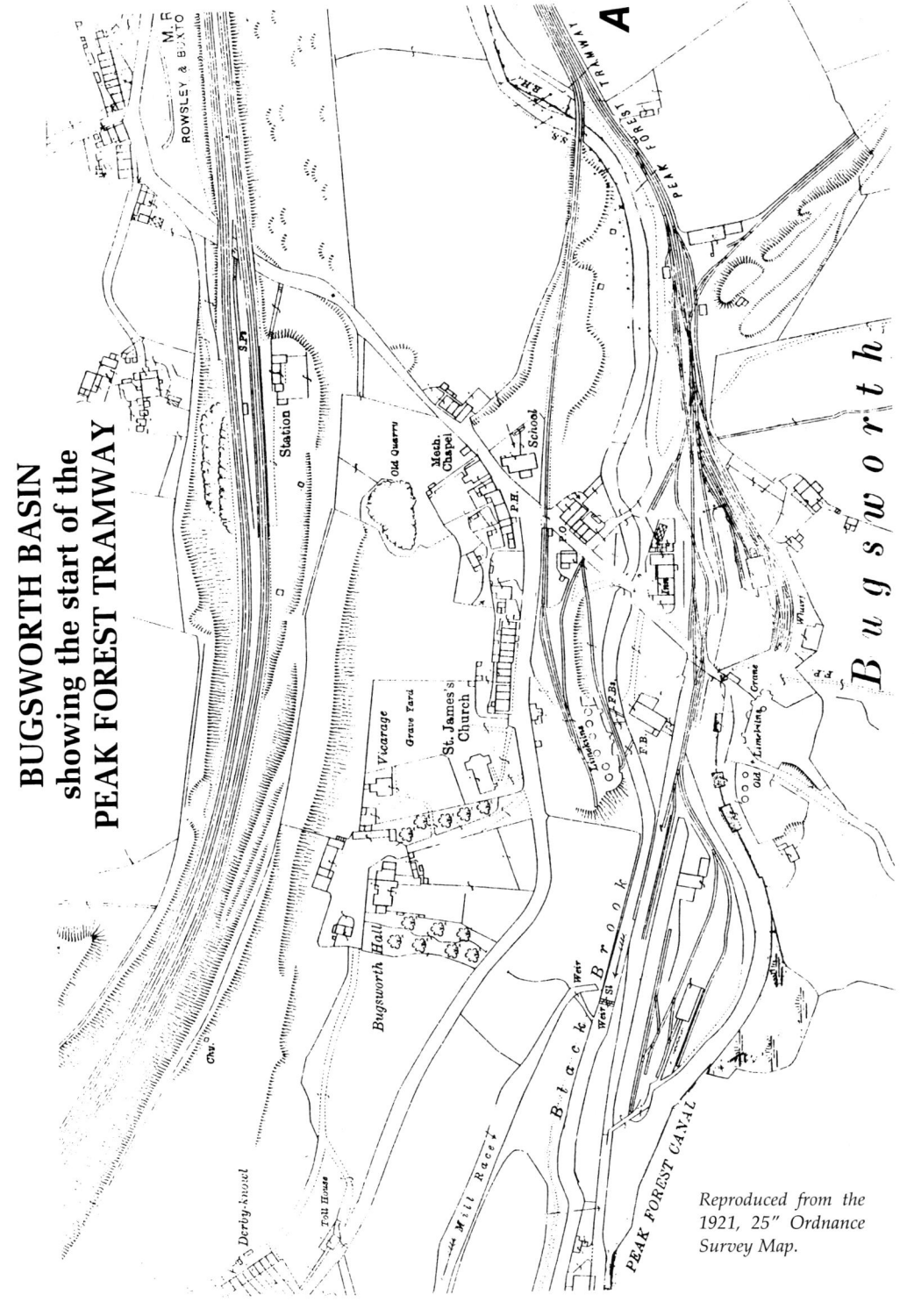

gauge lock and the amount of free board being measured; this measurement was then checked against tables held in the canal office. The boat then proceeded into one of the various arms of the canal according to the nature of the goods it was carrying.

The wharfage consisted originally of one basin, known nowadays as the "Upper Basin", constructed when the canal was built and probably completed in 1797; this basin included an area in which boats could be turned. As trade increased it was necessary to provide additional accommodation and further basins or wharfs were added during the next fifty years or so.

From the canal office with its row of windows facing the basin, we arrive at the gauging lock, ahead lies the canal to Ashton. It should be stressed that this is the main line and that the Whaley Bridge to Bridgemont section is only a branch arm. Turn round and we can see the entrance basin, the tow path, on the north side of the canal since Bridgemont, continues straight on into the lower basin, with its high wall. From here tramway trucks were unloaded, whilst on the opposite side a stone crusher had been erected by the Manchester, Sheffield and Lincolnshire Railway Company in 1860 to provide ballast for railway construction. Recently (1986) restoration work on the side of this arm has revealed the stone sleepers and a few sections of rail still in situ.

A bridge was constructed at the western end of the lower basin to allow horses to carry on past the lower basin arm with its "dry dock" to the middle and upper basins. This was known as the "island" by the old Great Central employees. The original bridge disappeared during the inter-war years, but now a replacement has been erected by the Manpower Services Commission working in association with the Inland Waterways Protection Society and ICI. Other work on the site includes rebuilding retaining walls and general site clearance.

Proceeding along the canal we pass the Wide, once a marsh where several streams entered the canal which had interrupted their route to the Blackbrook. This area has now been landscaped in conjunction with the Chapel-en-le-Frith and Whaley Bridge by-pass which is only a few yards away from the south side of the canal at this point. From the many rings on the bank side that once were used to tie up the boats, it seems that this must have been a busy area. But the tramway does not appear to have served this wharf, so it could be that empty boats awaiting their turn to load breasted up here.

The canal now enters the middle basin with lime sheds on the south bank; here boats could load, under cover, lime produced by the massive banks of kilns that once stood on either side of the road to Barren Clough.

According to the ICI records, the kilns at Bugsworth date from 1800 and 15 kilns in all were erected. In 1891 William Pitt Dixon who owned them joined in the merger of lime firms in the area to form the Buxton Lime Firms. The new firm had some 90 kilns scattered over the Peak District including a "Hoffman" kiln at Harpur Hill; several sets of kilns were closed, including those at Bugsworth and the ones by the top lock at Marple.

Most of the kilns have gone, some on the eastern side being removed in the 1950s to widen the road and recently several more to make way for the realignment of the road to enable it to cross the by-pass. Enough remains of two kilns, however, for us to visualize what they looked like in their heyday. If you look across from the tow path immediately before the middle basin arm you can see the "eye" of one of the kilns. When in use, stone and coal in alternate layers was tipped in the top of the kilns and hot lime "drawn-off" at the bottom through the eye, via the square opening.

It is now necessary to proceed under the road and we enter the middle basin arm, which was built to serve the lime kilns erected on the north side of the Blackbrook. This area has been tidied up by the IWPS. Leave the arm via the steps and proceed towards the road, noting on your right where the tramway once ran on a staging above the wharf. As you reach the road, a section of rail remains.

Over the road we reach the upper basin, on its north side more facilities for unloading stone, where the great "tippler" wheels once stood. These were wheels some 14 ft in diameter, and Farey gives the following description of their use:

> ... by means of a moveable Crane and Tippling Machine which runs on an outer Railway laid on purpose behind that on which loaded trams approach the sides of the boats. A man by means of a Winch-handle and chain which he fastens to the head of the tram, first turned round, on a turning plate, with its tail towards the Boat, shoots out its content.

This basin has two arms, one on the north-east corner, used for stone and the other parallel to it on the south side, the latter covered by a warehouse and interchange. On the south-west side, the remains of a crane can be seen; cranes of an identical pattern were situated on most of the wharfs along the Peak Forest Canal and were capable of transferring a loaded body off a wagon on to a boat or vice versa. This system was used for several years when the canal was first opened and the locks at Marple were not yet in use. Later the system was used for the carriage of salt to a bonded warehouse in Chapel-en-le-Frith and for other products.

THE ROUTE DESCRIBED

Bugsworth. The line in the foreground leads to the top of the lime kilns. The building on the right is the weighbridge for checking the amount of stone supplied.

Author's collection

The elevated branch line that lead to the lime kilns. The photograph is taken near the point where the branch joined the main tramway. The leftmost bridge crosses the Blackbrook the other took it over a farm track.

Ruth M. Brace

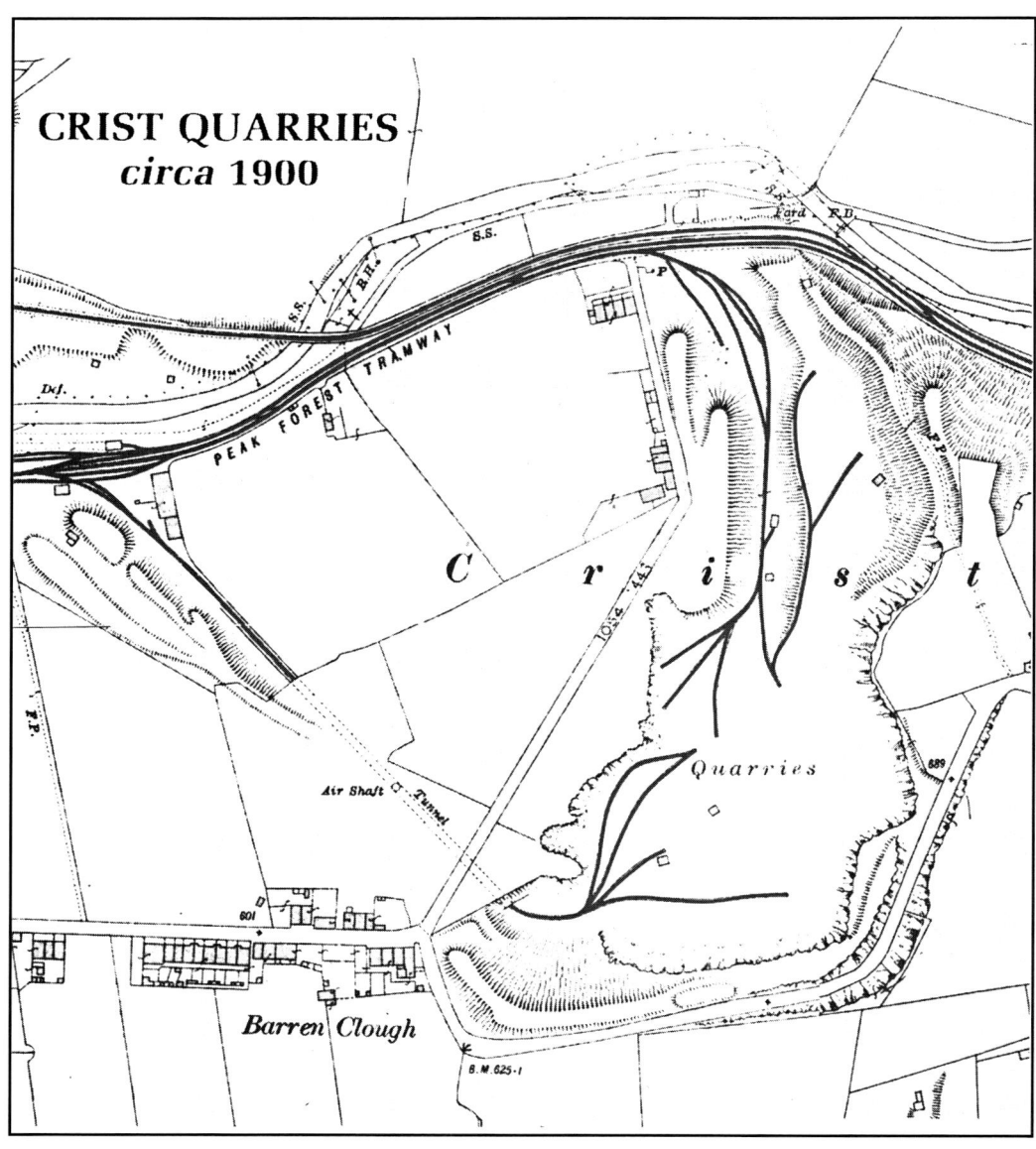

Crist Quarries

In making the tramway the company cut into a bed of hard gritstone about half a mile from the canal. It was realized that this was a good wearing stone and the company purchased additional land in Barren Clough. Stone blocks for the tramway were produced here as well as stone to build the wharfs, warehouses, bridges, locks etc., that were required along the canal.

An additional advantage of the stone was that when it was used for paving sets, it combined hard wear with the property of never becoming slippery after wearing smooth, and thus it was in special demand for stables and for paving streets in hilly districts.

It was soon necessary to extend the quarry and the road from Barren Clough to Eccles was diverted, at the same time a tunnel was built from the lower levels of the quarry to a point mid-way between the existing entrance and the basin: This tunnel was completely lost when the by-pass was built, as was the quarry itself, the spoil from the by-pass being tipped there and then landscaped. This had one mixed blessing in that it tidied up the quarry which had been the scene of much fly tipping since its closure around the 1920s.

Stone from the quarry was also used by the Manchester, Sheffield and Lincolnshire Railway and also by its successor the Great Central to build stations, bridges and docks.

Crist Quarry *c.*1905, home of the finest freestone on the Great Central Railway network!
Author's collection

The Bottom Line

The first part of this section, that from Bugsworth to Chapel Milton, was originally intended to be a canal, and in the Act provision was made for a reservoir to be built in the "Wash" to provide water for the flight of locks that would have been required. When the land for this section was acquired, several agreements were made with the land owners whereby they could have the fishing rights.

Outram, following the grant of the Act, re-surveyed the line and decided that the tramway should start from Bugsworth and not Chapel Milton or Chapel Milltown as it was sometimes called, and thus avoided the expense of both locks and reservoir.

From Bugsworth to the foot of the Plane at Chapel-en-le-Frith, there is a rise of 206 ft in a little over three miles and the line was constructed with a steady rising gradient. Leaving Bugsworth and Crist Quarry the Blackbrook is on our left and a branch which originally led to the top of the kilns crossed the brook to join the main line, where a crossover was installed. Shortly the base of a gasometer which once served the area can be seen at the bottom of the lane leading to Rosybank. The by-pass is now on our right having cut through the spur from Crist, leaving high above us several houses at Crist Knoll. Soon we reach Whitehall Works, extended many times since it was a paper mill which boasted that it made the largest lengths of paper in England. Crossovers and a siding were provided for the works and coal was delivered to the dyeworks which followed on the site of the paper mill, this having closed several years earlier.

The line now veers away from the by-pass and we cross the lane from Whitehough to Leaden Knowle, and once more we are by the side of the brook. In fact we are now on a vast earthwork that was built by Outram, in places 40 ft wide and up to 20 ft high and upwards of 100 yds in length.

At the end of the earthworks it crosses Green Lane where, local legend has it, one day a wagoner had just released his last brake on a train of wagons as he passed the Forge Mill, to enable him to maintain his speed through the almost level stretch through Whitehough. As he approached Green Lane he saw a cart drawn by a frisky horse making its way down the lane, but too late for the driver of the cart to take avoiding action. The horse hit the train, the cart flew up in the air and a well-known local doctor found himself on the top of a loaded truck on its way to Bugsworth.

Along the valley next comes the Forge Mill, once an iron bar slitting mill, then a paper mill, later used for bleaching and now part of Dorma;

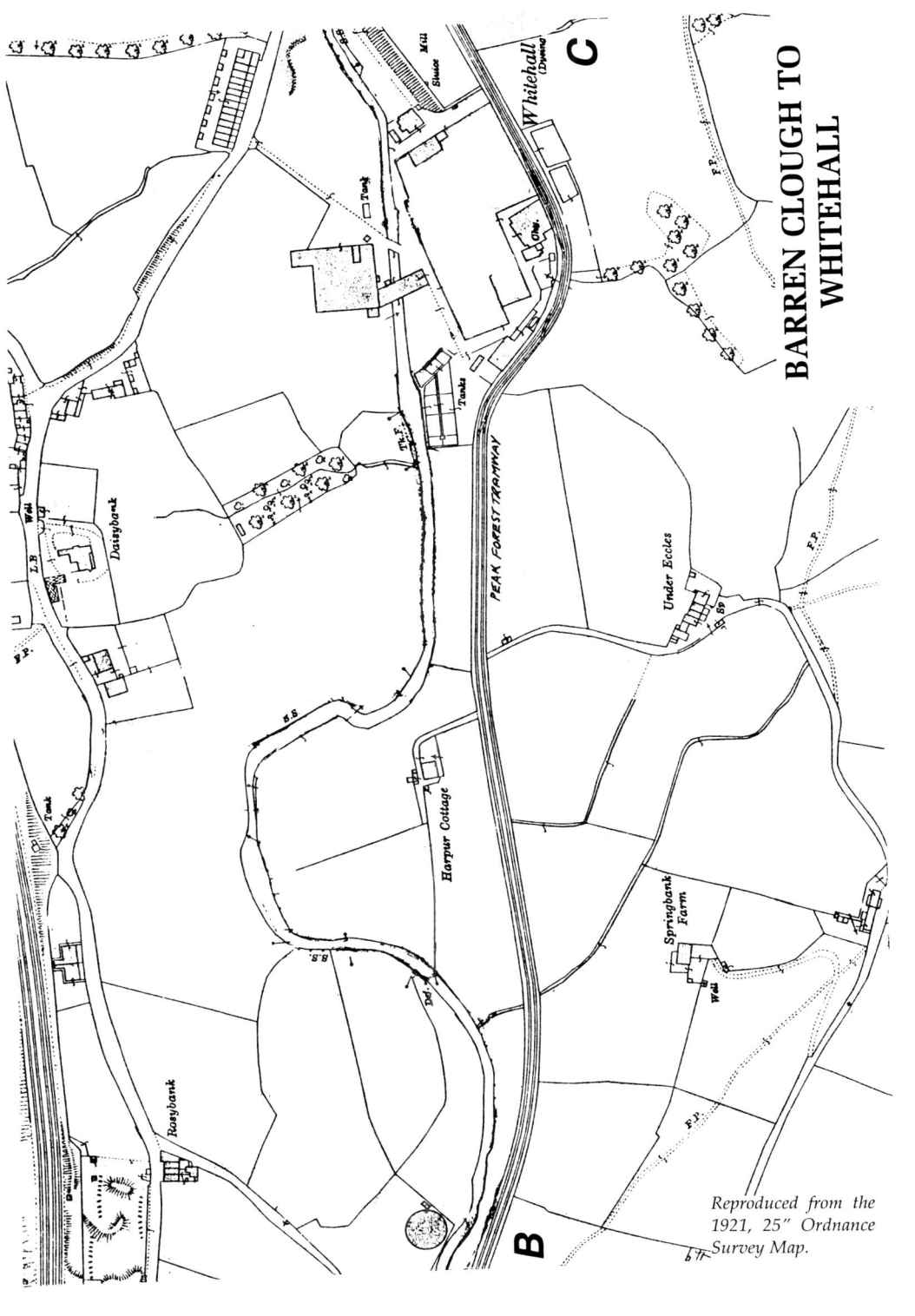

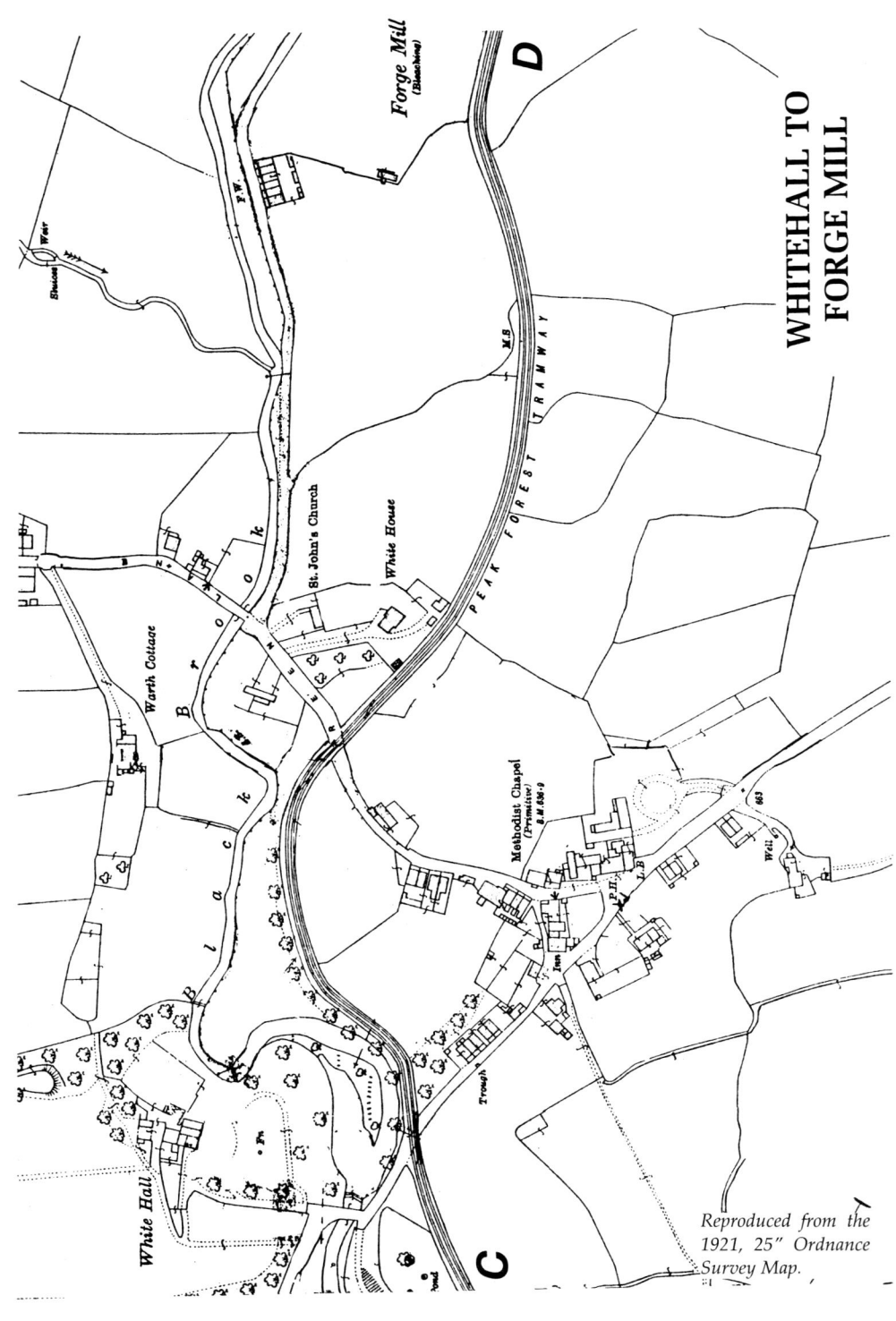

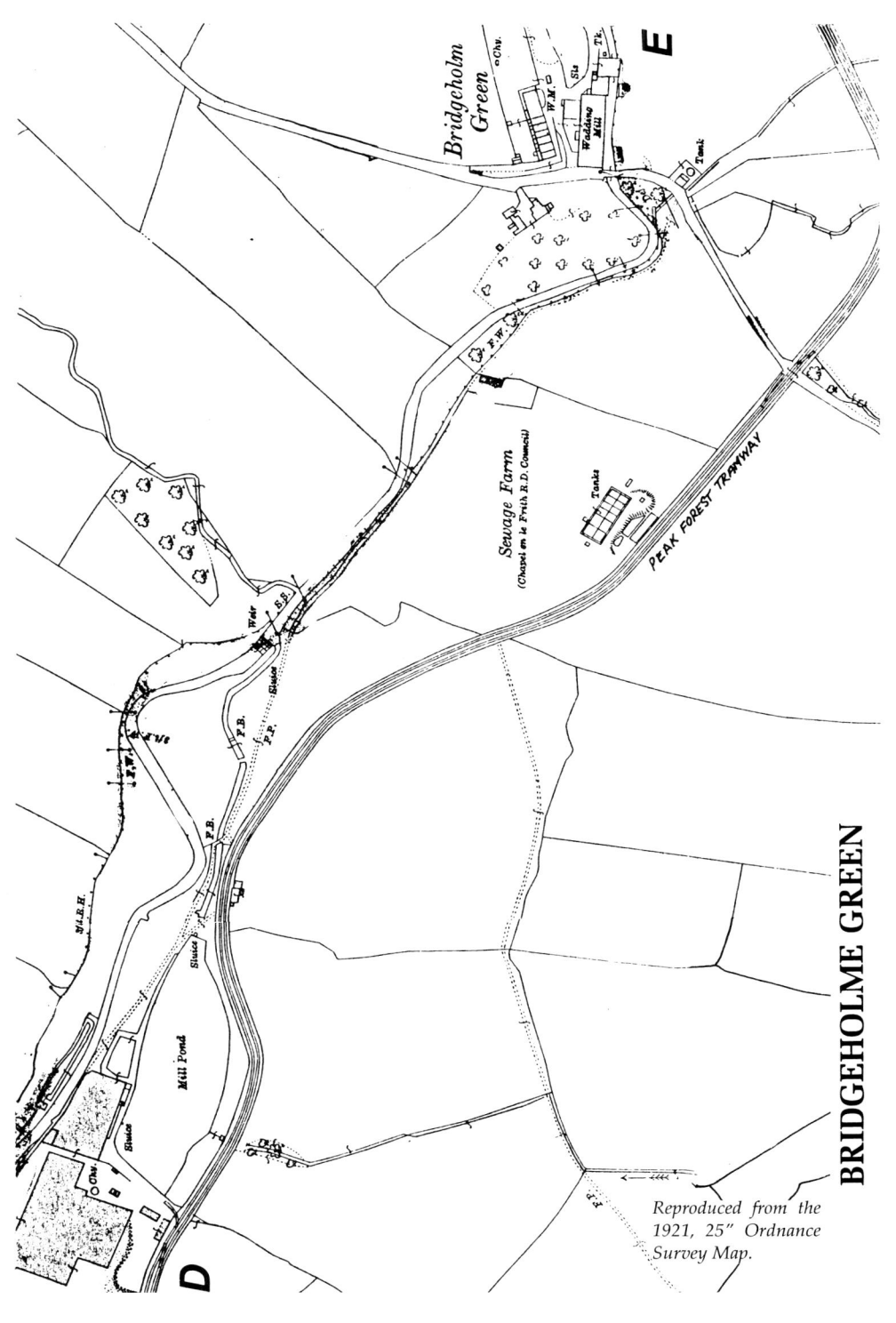

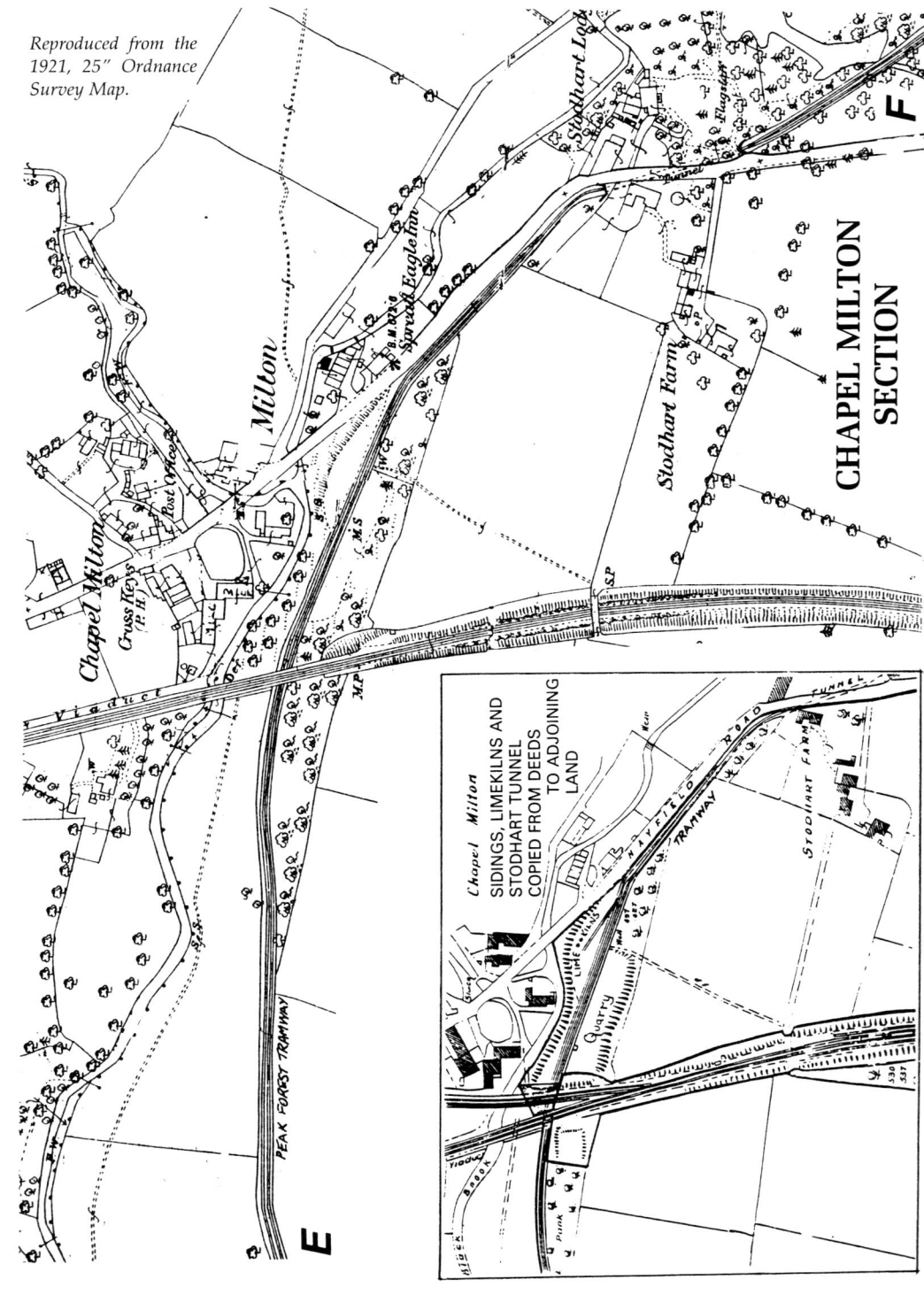

Reproduced from the 1921, 25" Ordnance Survey Map.

CHAPEL MILTON SECTION

SIDINGS, LIMEKILNS AND STODHART TUNNEL COPIED FROM DEEDS TO ADJOINING LAND

again crossovers and a siding were provided, mainly for coal but also bleached cloth was sent down from the Forge to another mill near Marple by the canal and tramway. Twisting and turning as it follows the contours, it reaches Charley Lane where crossovers and a siding were provided to serve the Wadding Mill a hundred yards or so away at Bridgeholm Green.

We have now joined once more the by-pass and over 100 yds of the track has been lost under the embankment that carries the by-pass. Here we have to leave the tramway and proceed down towards Bridgeholm Green and take the footpath through the mill yard which will take us to Chapel Milton. The section of the tramway that still remains between Charley Lane and Chapel Milton is now used as a high speed test track by the Ferodo Brake and Clutch Lining Company as far as the viaducts, and then a short length is occupied by filter beds. On reaching the main road turn right and on the footbridge pause to look at the proposed terminus of the canal. If it had been built would the Midland Railway have built the viaducts or would there have been many mills in the area?

A wharf was constructed here by the canal company for the benefit of goods, mainly lime for Hayfield. During the 1830s a lime kiln was constructed and in March 1860 the MS&LR Co. purchased it from John

Ferodo test track 2006, constructed on the trackbed of the tramway.   *Ruth M. Brace*

and Thomas Brocklehurst for £50. It was out of use by the late 1880s. The siding was used extensively whilst the railway to serve Stockport Corporation's Kinder Reservoir was under construction. Lime and some limestone were loaded onto carts and carried to Hayfield; when the Reservoir Railway was ready, the materials were sent by rail under a joint agreement between the Midland and Great Central Railway Cos.

Proceed up the hill and the footpath under the by-pass viaduct is on the bed of the tramway for a few yards. Originally it turned and entered a tunnel under the road emerging on the far side of Stodhart House Drive. The tunnel was severed when road improvements were carried out in 1949 and the southern entrance can still be seen.

South portal of the Stodart tunnel. *Ruth M. Brace*

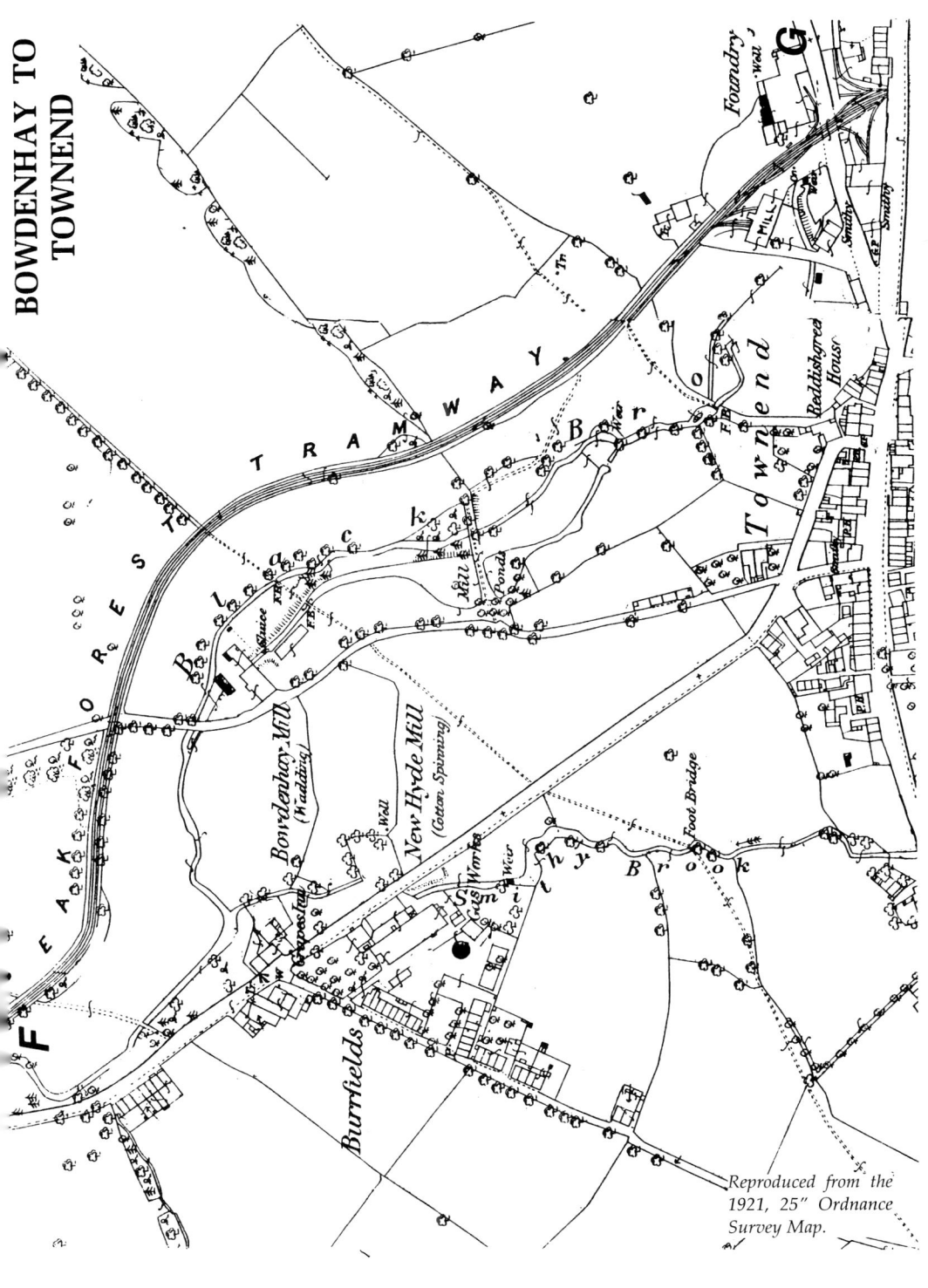

Whilst the line was originally laid with a single track with, we presume, crossing places or loops, the trackbed was built to take double tracks. By 1803 traffic was such, that the laying of a second line was authorized. However the tunnel at Stodhart was only built to take a single line and was never widened out, (recent examinations of the rocks in the area show very hard beds of rock which would have been very difficult to cut through). This caused hold-ups at both ends and consequently wagoners dis-obeyed the safety regulations laid down by the canal company in respect of the single line working through the tunnel.

One of the first recorded railway accidents happened here in the first decade of the 19th century when six loaded wagons which were being pulled up the line to Chapel broke away from the horse teams, ran back and in the tunnel hit a two-horse team hauling empty wagons. The two horses were killed and the nipper in charge was seriously injured; it was several days before the tunnel was cleared and the line re-opened. At the subsequent inquiry held by the canal company the blame was attached to the nipper leading the second horse team as he had failed to observe the correct interval before entering the tunnel.

Shortly after emerging from the tunnel the line crossed once more the Blackbrook on a magnificent bridge which unfortunately fell during the construction of a link road to the by-pass. Passing the rear of a car park serving the huge Ferodo Works, the line crossed Bowden Lane where there was once a siding to serve the tanning pits nearby (tanning having been carried out in the area for almost two centuries), the line continued on to the site of the marshalling yards and wharf at Town End. Parts of this section from Bowden Lane are private and so we proceed up Bowden Lane, turning left on Hayfield Road, left again on Buxton Road and in 50 yds left again in Castleton Road. Here we once more meet up with the tramway where it crossed the road into the old wharf (now a part of the County Council Yard) before it dipped under Buxton Road to the foot of the Plane.

The area at Town End was a very busy area, not only was it a marshalling point for the "Plane" but it was also a wharf with warehouses and a branch to adjoining mills. Besides the warehouses, buildings accommodated blacksmiths, masons, carpenters, nailmakers, farmers and various other trades who helped to maintain the tramway.

Before the opening of the two main line railways into Chapel-en-le-Frith, there was a large mixed goods traffic by canal and tramway from Manchester to Chapel-en-le-Frith for the towns, villages and works in the Peak District. From here there was a regular service of carrier carts to Buxton, Calver, Tideswell etc., as well as a regular service of night boats between Bugsworth and Manchester. One local carrier by water was James Walton whose brother Charles acted as agent at Chapel.

"Bottom of the plane", showing loaded wagons descending taken from the short tunnel under Buxton Road. *Author's collection*

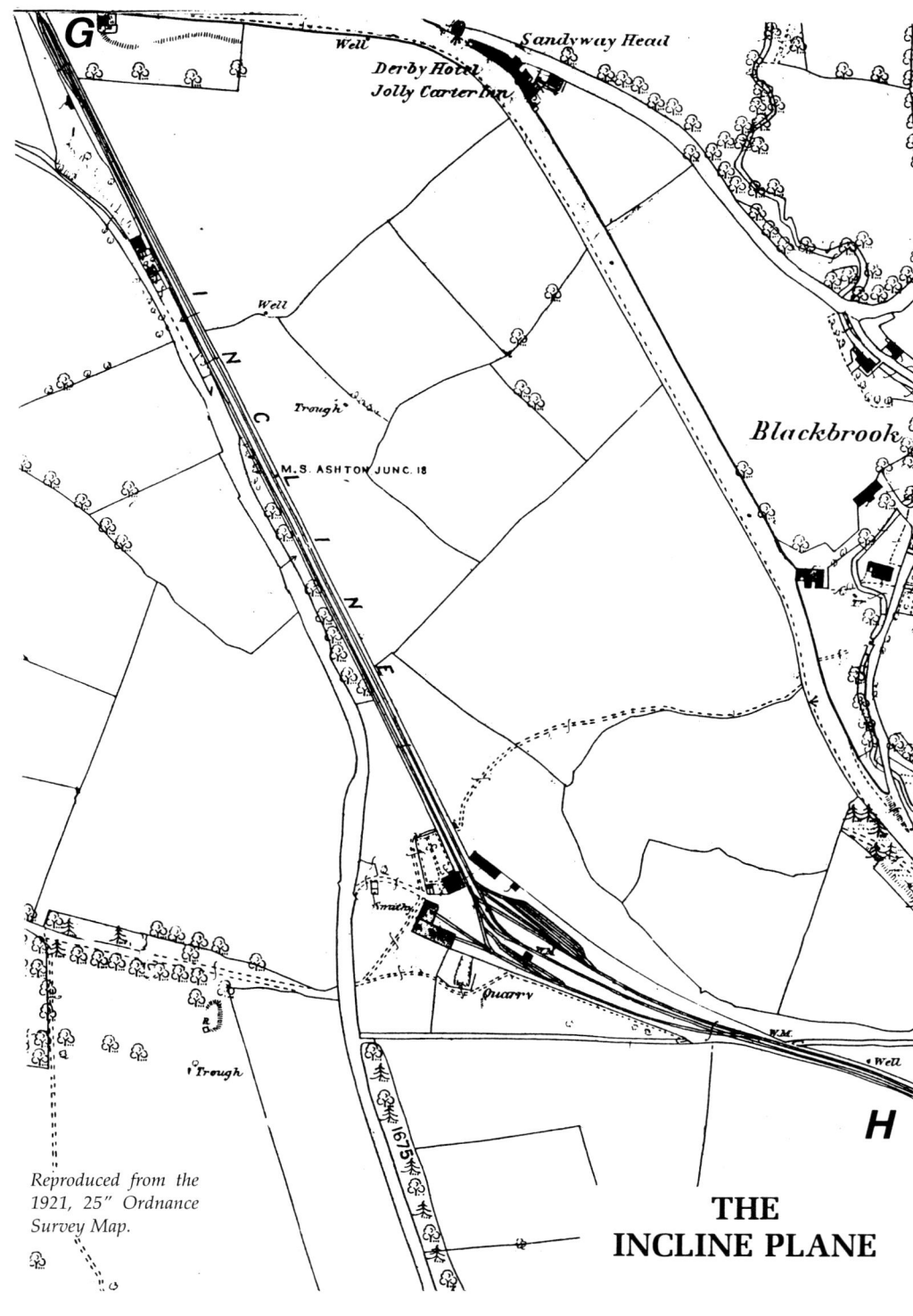

The Plane

The Plane was 512 yds long and rose 209 ft; this statement is in itself a tribute to the vision of the surveyor who originally planned the route and to the engineers and workmen who built it and later worked it for many years.

But let us look at the Plane in more detail; after passing under Buxton Road, the line branched out to give an extra line for storage of wagons awaiting their turn to ascend. This area was designed so that a loaded wagon, when at the bottom of the plane, would require a horse to pull it up a slight gradient and then under the road to the Town End Yards.

Originally a hemp rope was installed to connect the wagons that were going up and down the Plane, however, it was not strong enough and was replaced by a patent twisted chain. Farey records that this was also too weak and in 1809 a new chain some 1075 yds in length was provided at a cost of £500. This chain had 5 inch links and was manufactured in Birmingham. Later the chain was replaced by an endless steel wire rope, 2 inches in diameter.

Left: The Brakeman's Tower. *Right*: Loaded wagons arriving at the top of the plane. Note the haulage ropes between the lines. *Author's collection*

As the weight of the load going down in all cases exceeded that of the load going up, a means of controlling the speed and stopping the wagons at the end of the run was needed, this was done by a brakesman in his tower at the top of the Plane.

The endless rope at the top of the Plane passes underground and takes a turn and a half round a horizontal brake wheel some 14 ft in diameter. The groove in which the rope ran was lined with shaped wood blocks, so fitted that the end of the grain was outwards so as to increase the grip of the rope.

Above the top flange of the groove, the rim of the wheel was flat and about 5 inches deep. It was encircled by a steel strap of the same depth lined with wood brake blocks, these too were set with the end grain facing the wheel rim. The strap was shackled at one end to the wheel pit, and the other was shackled below the fulcrum of a strong steel lever some 15 ft in length which passed through the floor of the brakesman's hut. The lever was pulled over by means of a block and chain mechanism and thus formed an effective and powerful brake.

The incline is not true, it has a curve in it and this was of assistance in controlling the speed of the descending wagons. As the wire rope weighed between 5 and 6 tons the friction in passing over the supports between the line was considerable. The steeper gradient at the top with a flatter gradient at the bottom meant that as the descending load went

Marshalling wagons in the yard at the "top of the plane". Note the low sided wagon and the non-standard high sided wagons. *Author's collection*

down and the ascending load came up, the weights became equalized near the middle of the run, (due to friction); when nearing the end, the differences in the gradients greatly assisted in stopping the wagons. Originally the cable support between the lines were blocks of wood, placed every 10 yards, but these were replaced in the 1870s by rollers manufactured at Gorton.

The cable was known to break on several occasions and, depending where it broke, either descending loaded wagons would hurtle down and pile up under the bridge, or, as occasionally happened, the cable break allowed the empty ascending trucks to run away; these were quite likely to jump and land on the road! During the First World War a cable broke, the wagons piled up and one wagon it was claimed, had jumped over the bridge into the yard on the far side.

One of the first signals used on any line was used to control the working of the Plane, being a disc mounted on a staff, one side of which was painted white and the other side red; it was controlled by the ganger (foreman) in charge of the men who worked the foot of the Plane. When the empty wagons had been attached to the cable, (originally when it was a chain, by threading a short thin chain through the loops of the endless chain) by means of a cable and screw clamp, the ganger checked that the men had moved away from the line and rotated the disc to signal to the brakesman that he was ready. The ganger at the top of the Plane would verbally advise that he was ready, then the brakesman would release the brake to allow traffic to proceed.

Stable at the "top of the plane", 2006.   *Ruth M. Brace*

The Top Section

At the top of the Plane we reach the site of the huge marshalling yard that once stood there, with its range of stables, blacksmiths' and joiners' workshops, and a house for the brakesman. Today the buildings still remain and most are in good condition, but their use has changed. Several public footpaths pass through the yard and one follows the route of the tramway for a short way.

From here it was a long steady pull to Dove Holes and the ruling gradient is fairly constant at 1 in 200 until it reaches its summit in Dove Holes Dale. A letter written by Mrs Grace Bennett, once a close friend of John Wesley, tells how when she was 80 years old (in September 1795), she walked to Barmoor Clough from her house at Stodhart Lodge to see the railway and watched men at work on the narrow ledge along the side of the Clough.

However we are unable to follow it all the way and we have to join the A6 as it climbs up the Clough to Barmour Turning. After approximately ½ mile, the line had a branch siding which ran over a sett, here coal was unloaded for a factory on the opposite side of the road and for the laundry which succeeded it. Crushed stone for road making was also unloaded here into carts that were waiting below.

In another ¼ mile we reach the site of a further siding which was used for various traffic to and from the area around Peak Forest, Sparrowpit and Tideswell. Provisions and animal feedstuffs, coal, farming implements and mining equipment were brought up from Bugsworth and descending traffic consisted of lead ore, pigs of lead and other minerals from the mines in the Peak Forest and Sparrowpit area. The economics of this traffic can best be illustrated by saying that in 1808 the wagon hire for carrying coals from Bugsworth Wharf to Dove Holes or for "down" traffic was 7*d*. per ton whereas the quoted rate to hire a one-horse cart and man was 5*s*. per day or for a two-horse team 8*s*. A horse team would have been hard pressed to complete more than one trip a day from the Clough to Bugsworth and would have only carried one ton per horse.

In an old account book, it is recorded that a new winding winch for a mine at Sparrowpit was purchased in Liverpool, the carriage charge by canal and tramway from there to Barmour Siding being £1 3*s*. But to take it the last mile and a half, using a farmer's cart, albeit including loading and unloading cost 14*s*. and six gallons of strong ale.

The line now swings away from the road and for many a year a dam to serve the mill separated the two; we now reach the point where a

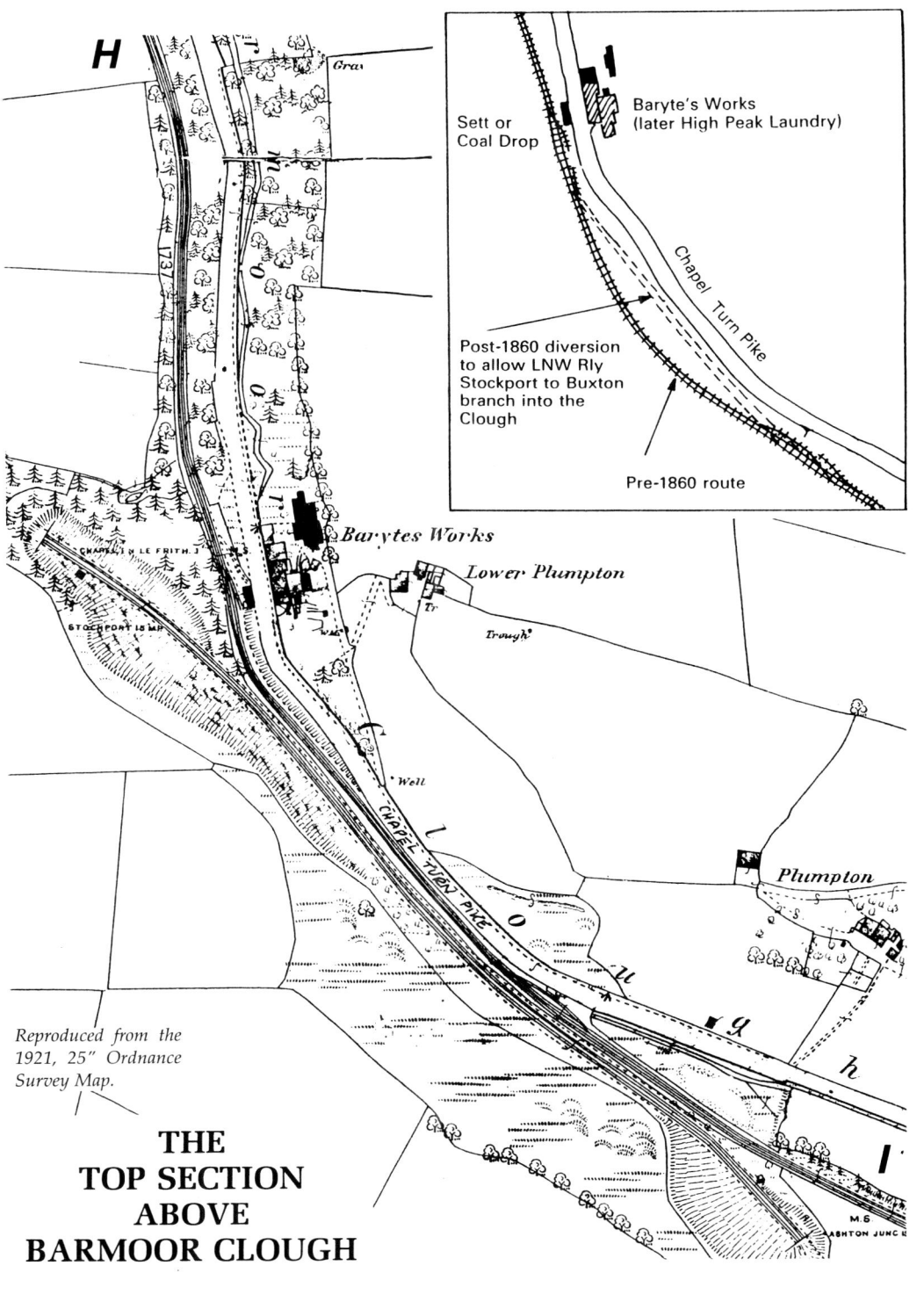

THE TOP SECTION ABOVE BARMOOR CLOUGH

Reproduced from the 1921, 25" Ordnance Survey Map.

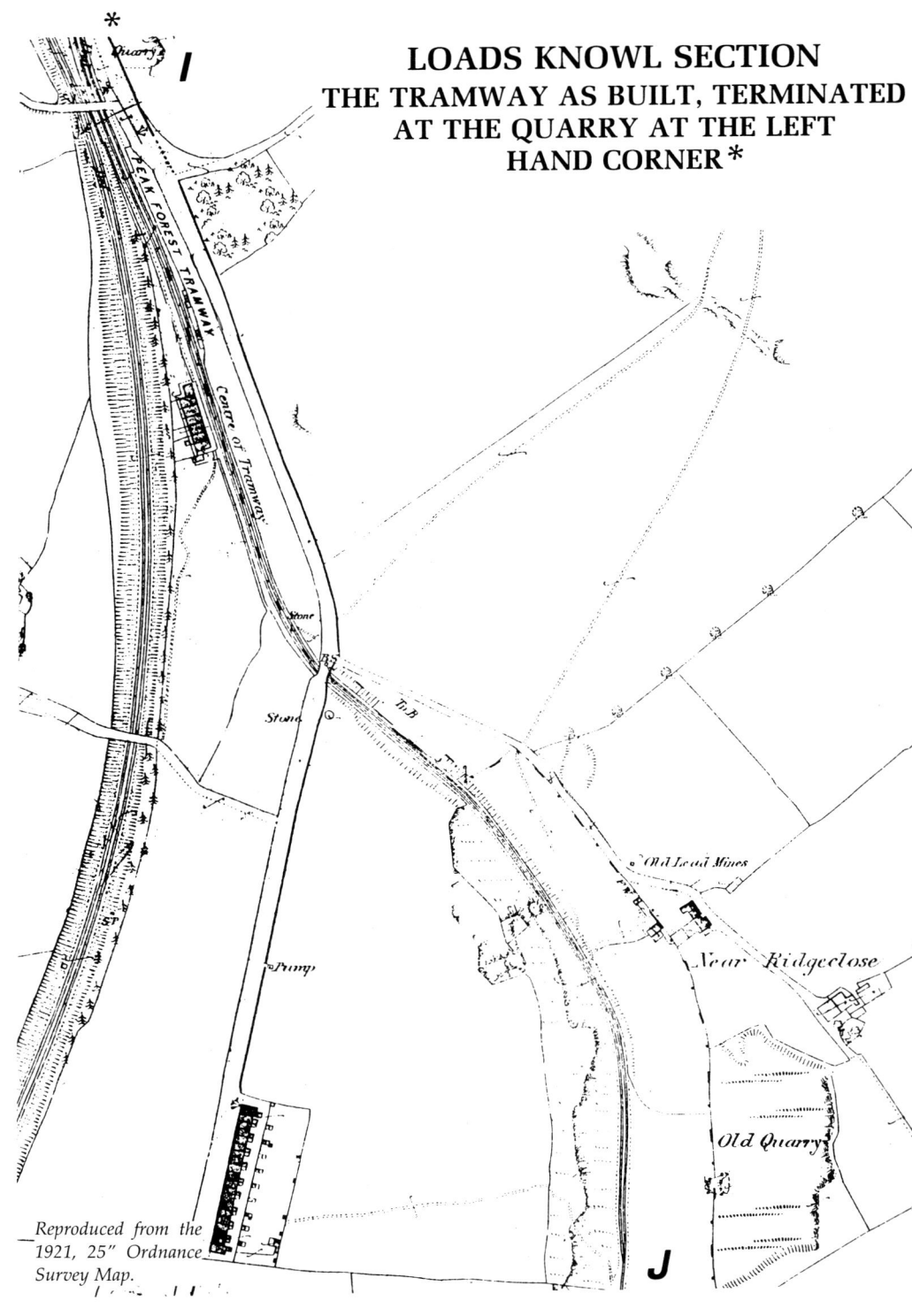

THE ROUTE DESCRIBED

Bridge over the tramway and the LNWR Buxton line. It was built, with the railway, in the early 1860s to replace a level crossing where the tramway was crossed by a minor road.

Ruth M. Brace

Looking towards Barmoor Clough from the A6 bridge over the tramway, near Dove Holes. The route of the tramway is marked by the boundary walls.

Ruth M. Brace

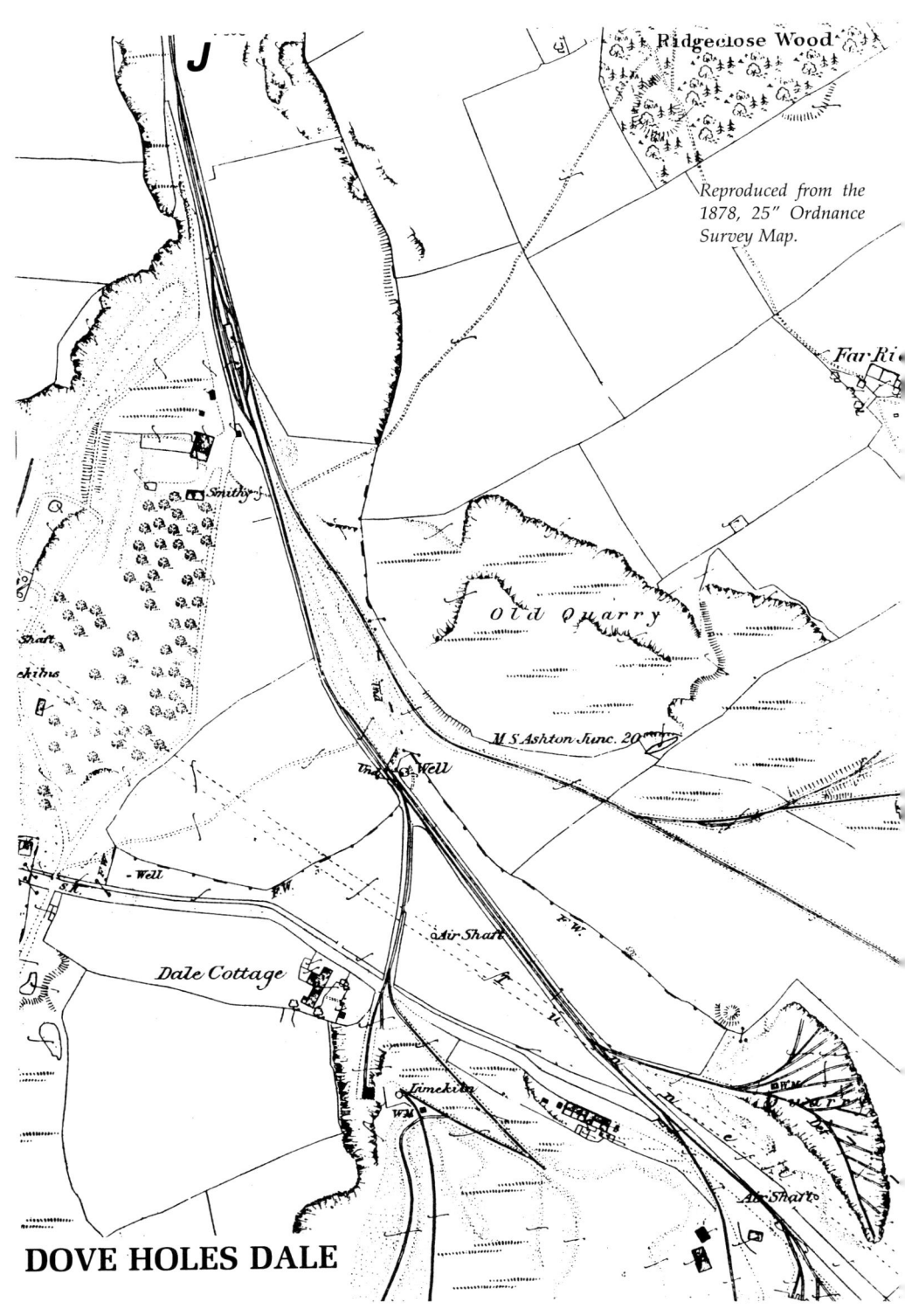

second incline took the line into Loads Know! Quarry. All we can see today of this quarry are a few outcrops of limestone and little remains of the incline which was only 33 yds in length, laid with a double track. Unlike the Chapel incline, a horse gin was used to draw up or lower trucks, this would no doubt be due to the lack of storage roads because empties were not always available when loaded wagons were ready for dispatch.

Within a few months it was evident that the supply of stone from Loads Know! was not sufficient to meet the demand and the line was extended into Dove Holes Dale. Leaving the Clough, the line ran along below the road to Buxton, this being constructed some five years or so after the tramway reached Dove Holes. A row of cottages was built alongside the tramway to house employees (Blackhole Cottages) and then the line swung under the Buxton Road (A6) into Gnat Hole at the head of Dove Holes Dale. The bridge as we see it today was a reconstruction by the County Council in 1935 when a road widening scheme took place. Although the tramway was by now closed, the rails were still in situ and the Council had to leave a bridge instead of filling in and forming an embankment. Here kilns were opened in 1808 by

Looking along the trackbed of the tramway to the bridge built in 1935 to carry the A6 over its remains."
Ruth M. Brace

Bed of the tramway passing limestone outcrops at Near Ridgeclose (*bottom right map page 48*).
Ruth M. Brace

George Potts, the stone being mainly supplied by Gisborne Quarry. The area was known as the Hallsteads and had been quarried for several decades before the tramway arrived.

Further down the Dale, the line branched out into the many quarries that either existed or which were opened once the tramway provided an easy method of transport. The canal company purchased a further quarry in the area which they worked until the 1870s and then sold to one of the canal carriers.

Holdness Quarry and two lime kilns were operated from 1860 by Joel Carrington, a lime merchant of Hollinwood near Oldham, and in 1879 he sold out to Samuel Taylor, the son of a tailor who had left Chapel in the 1830s to set up in Runcorn as a merchant and carrier of coal and lime by canal boat. This became S. Taylor Frith and Co. Ltd, in 1905.

A branch to the quarries on the southern side of the Dale crossed the road on a brick bridge which soon came to be known as Red Bridge, it was such a contrast to the lime and limestone, the dust from which whitened the landscape and gave it in parts a lunar look. Lime sheds were erected; these were low buildings where loaded wagons were stored overnight as lime, when it gets wet, starts to heat up and change into quick lime, this process is known as "falling".

Leaving the lime sheds the wagon would be hauled to the summit of the line where the ganger would take charge, leaving the nipper to follow with the horses. Haulage to the various quarries not owned by the canal company was provided by private teams, one family, the Marchington's of Dove Holes had teams for over fifty years and a Sam Marchington was still providing haulage for the Perseverance quarry and lime kilns right up to the closure of the line.

Most of the limestone quarries and lime works in the area merged in 1891 to form the "Buxton Lime Firms" leaving only 4 independent firms, one of which was S. Taylor and another the Manchester Sheffield and Lincolnshire Railway Co. which still operated the New Lime Quarry. In 1923 the lease of the "New Line" was sold to S. Taylor Frith & Co.

Quarries connected to the tramway in Dove Holes Dale according to L. Jackson in the *Buxton Lime Trade* were:

Holderness, Newline, Gisborne, Heathcote, Perseverance and Wainwright.

Three-holed sleeper block in situ at Dove Holes Dale.   *Ruth M. Brace*

THE ROUTE DESCRIBED

The top end of the line, where it branched into the various networks serving the local quarries.
Ruth. M Brace

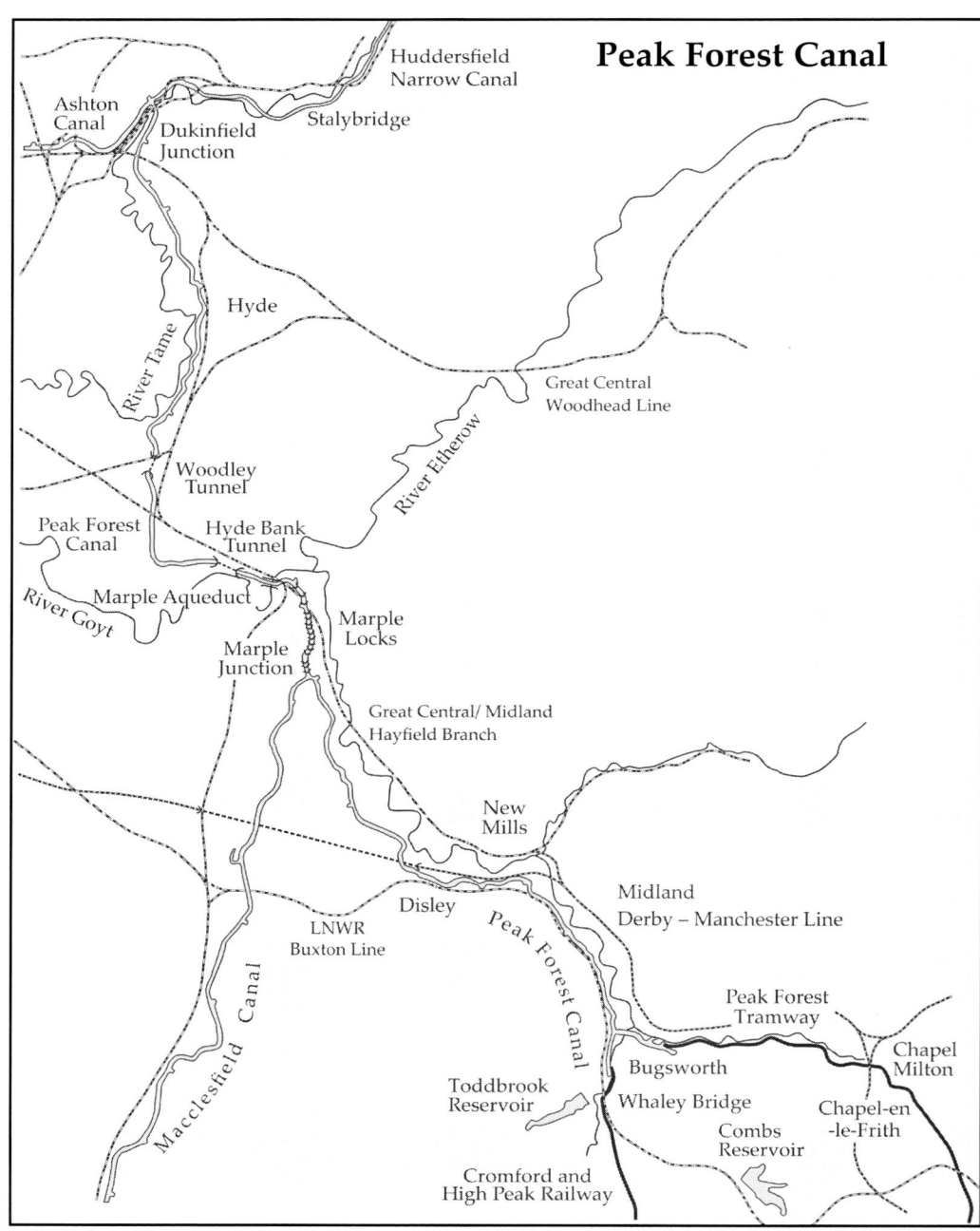

Chapter Five

The Peak Forest Canal

History

By the 1794 Act which was for

> making and maintaining a navigable Canal from out of the Canal Navigation from Manchester to or near Ashton-under-Lyme and Oldham in the County Palatine of Lancaster, at the intended Aqueduct Bridge in Duckinfield in the County of Chester to or near to Chapel Milton, in the County of Derby and a communication by Railways or stone roads from thence to Loads Knowle, within Peak Forest in the said County of Derby and a branch from out of the said intended Canal to Whaley Bridge in the said County of Cheshire

the company was to be known as the "Company of Proprietors of the Peak Forest Canal" and was authorized to raise £90,000 in shares of £100 and up to £6,000 on mortgage of the rates.

However this sum fell short of the monies needed, so in May 1800 a second Act was obtained (39 & 40 Geo III Cap 38; 30th May, 1800); this Act allowed for an increase in the capital. As the original sums authorized by the first Act had been expended on building the section

A laden narrowboad moored by the canal company offices at Bugsworth Basin.
Author's collection

from Ashton to the foot of the locks and the top sections of the canal, work was authorized in 1801 for the construction of the Marple flight of locks and a reservoir at Combs, near Chapel-en-le-Frith.

Little money was forthcoming and it was not until 1803 when Samuel Oldknow and Richard Arkwright provided the increased capital in return for toll concessions that the work could commence. Once the monies were to hand Jones and Fox were appointed contractors for the locks, they had completed them by late 1805 and the locks were in use in January 1806. The reservoir at Combs had been started and water fed the canal via a leet to Whaley Bridge.

The land for the reservoir was partly purchased by the canal company and partly by Oldknow and Arkwright; later (9th November, 1809) this land was conveyed to the Peak Forest Canal Co. During this period the Peak Forest Canal Co., and the Ashton Canal Co., placed the canals under a Joint Manager, this arrangement was later extended by the appointment of a Joint Engineer.

The Macclesfield Canal which was to link the Peak Forest Canal to the Trent and Mersey Canal was authorized in April 1826 after some thirty years of talks, surveys and unconcerted action by the canal companies and the towns in between.

Oldknow and Arkwright were subscribers to the Macclesfield Canal which was built on a line surveyed by Telford under the direction of William Crosley. The canal was opened on 9th November, 1831, it is 26

The junction of the Peak Forest Canal with the Macclesfield Canal at the head of the Marple Flight of locks. *Author's collection*

A Sunday school outing *c.*1905, from Bugsworth to Marple seen here at Rawton Walls Bridge. *Author's collection*

The canal company repair base at Marple on the Macclesfield Canal. The Peak Forest Canal lies through the bridge. *Author's collection*

Two fine views of icebreakers on the Peak Forest Canal seen between the south end of Hyde Bank tunnel and the aqueduct. *Courtesy of Mrs J. Bailey*

miles in length and joins the summit level of the Peak Forest Canal to the summit level of the Hall Green branch of the Trent and Mersey Canal with only 12 locks in a staircase at Bosley.

Once the Macclesfield Canal was under construction, the Peak Forest Canal Co., decided to build a second reservoir, and this was constructed on the Todd Brook at Whaley Bridge and was completed by the mid-1830s.

After the Manchester, Sheffield and Lincolnshire Railway Co., had acquired the Ashton and Peak Forest Canal in 1845 and the Macclesfield in 1847, the three canals were jointly administered. The system was prosperous as the canals provided the MS&L Railway with a share of the rich limestone traffic of the Peak and it was not until the railway amalgamation of 1923 that the traffic began to fall away.

After 1923 tolls were increased, water traffic fell away, the Peak Forest Tramway was closed, the Bugsworth to Bridgemont section of the canal fell into disuse and was derelict by the late 1930s.

Marple aqueduct in January 1962 suffered considerable damage, and had it not been for the support of the Local Authorities, and the various voluntary canal societies, the aqueduct would have been demolished. The Local Authorities contributed some £50,000 towards the repairs which were completed in May 1964, and since then the structure has been officially designated as of outstanding historic and archaeological interest.

The aqueduct in 1960 after the collapse of part of the sidewall. *Author's collection*

(39)—GREAT CENTRAL RAILWAY—continued.

(i) *Types of vessels using the navigation.*

Narrow boats. Steam haulage is not allowed.

Peak Forest Canal.

(c) *Distance Table.*

Main Line (No. 39b1).

	Miles.	Fur.
Dukinfield Junction, junction with Ashton Canal (No. 39a1), to—		
G. C. Railway Bridge (Stalybridge Branch)..	–	$0\frac{1}{2}$
Dukinfield Gas Works ..	–	$2\frac{2}{3}$
G. C. Railway Bridge (Main Line)	–	5
Dog Lane Bridge	–	$6\frac{1}{4}$
Dukinfield Hall or Well Bridge	1	$0\frac{1}{2}$
Newton Hall Bridge	1	$5\frac{1}{4}$
Hyde, Bowler's or Throstle Bank Bridge	2	1
Hyde, Company's Warehouse and Wharves ..	2	2
Hyde Gas Works	2	4
Apethorne Aqueduct over road	3	$1\frac{1}{2}$
Woodley Bridge	4	2
Cheshire Lines Railway Bridge (Woodley and Stockport Line) and North end of Woodley Tunnel	4	3
Sheffield and Midland Railway Bridge (Romiley and Stockport Line)	5	0
Hatherlow, Holehouse Fold Bridge	5	$2\frac{1}{2}$
Hatherlow Aqueduct over road	5	4
Chadkirk Aqueduct over road	5	$5\frac{1}{2}$
North end of Hyde Bank Tunnel	6	$1\frac{1}{2}$
South end of Hyde Bank Tunnel	6	3
Occupation Bridge	6	4
Marple, Aqueduct over River Etherow	6	$6\frac{1}{2}$
Marple, Sheffield and Midland Railway Bridge (Marple and Romiley Line), and Company's Wharf and Workshops	6	7
Lock No. 1, Marple	7	0
Marple, Sheffield and Midland Railway Tunnel under canal (Marple and Romiley Line) ..	7	2
Marple, Company's Warehouse and Wharf ..	7	$5\frac{1}{2}$
Marple, Head of Lock No. 16, and junction with Macclesfield Canal, Main Line (No. 39c1)	8	$0\frac{1}{2}$
Rawton Walls Bridge ..	9	1
Moore's or Turf Swivel Bridge ..	9	5
Stanley Hall, or Kicker's Bridge	9	$7\frac{1}{2}$
Woodend Swivel Bridge	10	$2\frac{1}{2}$
Higgins Clough or Shaly Knowl Swivel Bridge	10	$4\frac{1}{2}$
Disley, Dryhurst Bridge	10	$7\frac{1}{2}$
Greens Hall Bridge	11	2
Wirksmoor, Company's Warehouse and Wharf	12	$0\frac{1}{2}$
Bankend Bridge	12	3
Carr or Mellor's Swivel Bridge ..	12	6
Aqueduct over Furness Brook	13	1

Extract from Bradshaw's Canals and Rivers for 1904.

THE PEAK FOREST CANAL

(39)—GREAT CENTRAL RAILWAY——*continued.*

	Miles.	Fur.
Dukinfield Junction, junction with Ashton Canal (No. 39a1), to (*continued*)—		
Bong's or Yeardsley Bank Swivel and Foot Bridges	13	3½
Greensdeep or Bugsworth new Road Bridge	13	5
Junction with Whaley Bridge Branch (No 39b2)	14	0
Aqueduct over River Goyt	14	1
Bugsworth, Chinley Road Bridge	14	5½
Bugsworth, Termination of Canal	14	6

Whaley Bridge Branch (No. 39b2).

Junction with Main Line (No. 39b1) to—		
Roots Wharf	–	1½
Whaley Bridge, Company's Warehouse and Wharf	–	3½

(*d*) *Locks.*

Main Line (No. 39b1).

1)
to Marple.
16.)

Rise from Dukinfield.

(*e*) *Maximum size of vessels that can use the navigation.*

Main Line (No. 39b1) and Whaley Bridge Branch (No. 39b2).

	Ft.	In.
Length	70	0
Width	7	0
Draught	3	3
Headroom	5	10

(*f*) *Tunnels.*

Main Line (No. 39b1).

Woodley—
Length	176 yards.	
Minimum height above water level	9	0¼
Minimum width at water level	9	3½

Towing-path through the tunnel.

Hyde Bank—
Length	308 yards.	
Minimum height above water level	6	8
Minimum width at water level	16	0

No towing-path boats "legged" or "shafted" through.

(*g*) *Towing-path.*

There is a towing-path throughout the canal and Whaley Bridge Branch, with the exception of Hyde Bank Tunnel.

(*i*) *Types of vessels using the navigation.*

Narrow boats. Steam haulage is not allowed.

Extract from Bradshaw's Canals and Rivers for 1904.

Unfortunately during this period the locks on the Peak Forest and the Ashton Canals were not maintained, and by 1964 they were unusable. Thus the Peak Forest was no longer a through route, only the upper section remained and in the 1968 Transport Act under part 11 of the twelfth schedule under "Cruising Waterways" we note "The Peak Forest Canal from the top of Marple Locks to Whaley Bridge". Since this Act work by several voluntary bodies had prevented the further decay of the canal and through their dedication, the locks of the lower Peak Forest and the Ashton Canals were restored to navigation in 1975. This was followed in 1976 by the cleaning and re-opening of the Rochdale Canal section which joined the Bridgewater and Ashton and completed the Cheshire ring.

Route

The canal commences in the Manchester, Ashton-under-Lyne and Oldham Canal (The Ashton Canal) west of Ashton Wharf, it then crosses the River Tame and flows along the eastern side of the Tame Valley, through Newton Wood, to Woodley. Here it enters a short tunnel and goes via Butterhouse Green, Bredbury Green and Chadkirk to Romiley, a second tunnel then takes the canal under Hyde Bank to Marple Dale. It then crosses the river Goyt by an impressive aqueduct to reach the foot of the Marple Locks; 16 locks lift the canal some 212 ft to Marple. At the head of the locks the Peak Forest Canal is joined by the Macclesfield Canal. The Canal then carries on a level via Disley and Newtown to Bugsworth, with a short branch to Whaley Bridge.

Lock No. 15 at Marple. Note the side chamber allowing water to be saved from the lock above. *Author's collection*

Tunnel dimensions:

		Ft	In.
Woodley Tunnel	Length	501	0
	Min. height	7	6
	Min. width	7	9

(There is a towpath through this tunnel.)

Hyde Bank Tunnel	Length	924	0
	Min. height	7	0
	Min. width	15	10

Tonnage rates authorized by the Act of 1794

	per ton per mile
For all limestone	1 ½d.
For all other stone and lime coal & other minerals	2d.
For all dung, clay, sand & gravel not passing a lock	1d.
Ditto passing a lock	2d.
For all timber, goods & wares, other merchandise & other articles,"matters & things not herein before particularized.	3d.

In 1808 the rates had fallen and were:
For coals and stone	1d.
Limestone Lime	1½d.

Wharf dues were 3d. per boat

The tunnel under the canal at the junction with the Whaley Bridge branch. This tunnel was used by horses towing boats, to and from the Whaley Bridge section.   *Author's collection*

The canal near Greensdeep during 1966. The lower Peak Forest Canal was closed and the top section classified as a "Cruiseway" with little or no maintenance.

Author's collection

The canal at Disley in 1966.   *Author's Collection*

Chapter Six

Conclusion

Can any Railway or Tramway expect to span three centuries without changing its mode of operation? Whilst the Canal, it was built to act as a feeder to, is still going strong in its fourth century.

The Peak Forest Canal and Tramway devised and built in the 18th century is still with us today, not in its original form or carrying out the tasks listed in the preamble to the Act of incorporation. The tramway closed in the 1920's having scorned mechanical power or converting its track from plate way to edge way, the track was lifted in the 1930's for scrap and parts of the route were sold off.

The Canal suffered once the tramway closed, depriving it of the stone traffic it was designed to carry. Silting occurred and little or no maintenance was carried out, during the Second World War. On the upper section from Marple to Whaley Bridge and Bugsworth, traffic was virtually nil, the only reason it was kept was that it supplied water to industry in Manchester. On nationalization ait was downgraded to "Remainder" status – a waterway for which no commercial or leisure use was envisaged.

The millions of tons of stone that use to travel down the tramway and canal from the Dove Holes, area are still going down the valley, mainly in special trains, some by road using the by-pass which runs near to the route of the tramway from Chapel-en-le-Frith to Bugsworth and then alongside side the canal to Bridgemont. When Outram surveyed the route for the canal and tramway, little did he realize that soon a Railway, a road and later a Bypass would follow his preferred route.

However people with vision have helped to restore the canal, various societies were formed to restore lengths or the flights of locks. Today we see the canal once more a working canal, although in the main the boats are pleasure boats and the few working boats are those that carry supplies for sale to them or work on the maintenance of the canal.

The work on the Bugsworth Arm commenced in earnest in the late 1960's, the Inland Waterways Protection Society under the direction of Bessie and John Bunker were the driving force, (Bessie and John were leading lights in the Industrial Archaeology field). To their credit the IWPS are still the main developers of the site. Hard work by many volunteers aided by 'Bodices Chariot' resulted in the canal as far as the stop lock being in water by the early 1970's, however it was short lived, the stretch of canal leaked. On checking local records it was found that in the 1950's, the local RDC had run a sewer along the valley and in so doing had made use of the bed of the canal. The contractors like most others at that time thought the canal was finished and did not reinstate the canal lining, so a leak or a series of leaks developed. Work progressed with the IWRS helped by the Canal Recovery Group and the Army on several occasions, repairing walls, trying to solve the problems with the canal puddle, until 1999 it was thought that the work was complete, the canal was to re-opened and the Mayor of the High Peak Authority officially re-opened the canal at Easter. Some 40 boats turned out and were moored at Bugsworth for the event. It was said in the local press that the day was dedicated to Bessie and John Bunker.

However once again the water in the basin was short lived, the damaged caused in laying the sewer still gave trouble and the new puddle clay could not cope. By October 1999, water loss had become unacceptable and closure to navigation had to follow. Stop planks once more being used at Bingswood pinch. As the opening had caused a great deal of interest, British Waterways, together with the IWPS, the High Peak Borough Council and English Heritage formed a partnership to solve the problem for once and all.

Boats in the main arm of the basin in 1999. *Author*

CONCLUSION

Relining the canal with concrete to prevent further leaks in 2004, where the middle arms of the basin join the canal (*above*) and in the north arm (*below*). The area to the left of the ladder is the site of the crusher (*see page 24*).   *Author*

The objectives were to make the area from the junction with the Whaley Bridge branch and the whole Bugsworth complex, safe not only for boats and their crews, but also to form a safe area for the general public who now see the area as a recreational area, as it is easily accessible to not only the local populace, but also from the surrounding towns. A full ground investigation and topographical survey was carried out, funded by British Waterways.

At a cost of £1.2 million the entrance basin, the lower basin arm and a section of the canal from near the Britannia Mill (destroyed by fire in 2005) to where it passes under the bypass, were relined with concrete on a membrane and the puddle reinstated.

Easter 2004 saw the second re-opening. This was a gala event and many boats were present as well a several tents with displays by various Canal Societies etc. The event was fully enjoyed by all those who attended.

A study of the route of the Tramway was carried out by Entec UK in association with Asken Ltd. in 2004 to assess the feasibility of reinstating the Peak Forest Tramway as a footpath, cycle path and horse-riding trail, with access for the mobility impaired. It wasn't possible to establish a trail along the entire length of the tramway, but the section between Bugsworth Basin and Chapel-en-le-Frith became the Peak Forest Tramway Trail in 2006. Regrettably, this has not been extended to the upper part of the route, and still, in 2021, there is no footpath to Dove Holes along the tramway.

The Inland Waterways Protection Society became the Bugsworth Basin Heritage Trust in 2014. Around the basin, bridges have been restored and fencing renewed, there are interpretation boards, as well as a small exhibition about the tramway and canal. The trust continues to work to improve the basin and preserve its history. This has included building a replica tramway wagon, much of it new but incorporaating parts. like the stub axles and wheels, recovered whilst clearing the the basin.

Over 50 years ago an intrepid band of waterways enthusiasts made a journey from near Ashton along the overgrown, weedy neglected canal, thro' the locks (how remains a mystery, even to those who took part) to the junction at Bridgemont. Part of the journey under power, the rest being hauled and man handled by many including several who were out for a walk along the tow-path, Little did they know that their epic journey, along a discarded (at least by the Authorities) canal would be a turning point, and would lead to the present day glories of the Peak Forest Canal & Basin.

Appendix One

Chronology of the Peak Forest Tramway and Canal

1791	Scheme proposed to link the Ashton and Cromford Canals. William Jessop to be Engineer and Benjamin Outram his full time assistant.
1793	Plan for canal and tramway deposited by T. Brown with the Clerk of the Peace for Derbyshire.
1793	(5th Dec.) A meeting to promote the canal was held at the Ram's Head Disley.
1794	(28th Mar.) Peak Forest Canal Company incorporated, Peak Forest Canal Act (34 Geo. 3c. 26.).
	(April) Work on the canal commenced.
	(5th June) First General Meeting of the "Peak Forest Canal Co."
	(July) Reported a hold-up due to a shortage of labour and work would recommence on cutting as soon as the corn harvest was in.
	(7th Aug.) Samuel Oldknow of Marple elected to the Committee.
1795	Following a re-survey of the route, construction of the tramway commenced.
	Fund established for the purpose of relieving the necessities of the families of workmen prevented from working by reason of sickness, or by any accident happening to them in the execution of the canal, the company to contribute 20 shillings (£1) per fortnight to the fund.
1799	Tramway in use and extension into Dove Holes Dale under construction. A proposal was made to use the route of the tramway as part of a railway from Bugsworth to Buxton with a branch to Sparrowpit.
	(28th Sept.) Plans deposited with the Clerk of the Peace for Derbyshire for a railway from the canal at Whaley Bridge to run via Horridge End, Cadster, Tunstead Milton, Marsh Hall. Marsh Green, Hollinknowle and the Eaves to join the Peak Forest Tramway at the top of the Plane (not built).
1800	(30th May) An Act to raise additional Capital sanctioned. (39 & 40 Geo 111 Cap 38). Abraham Rees in Cyclopaedia states the works on the tramway were now complete.
1802	Shares quoted at £110 each on the London Stock Exchange.
	The company purchased part of the Kirke family estate in Dove Holes.
1803	The tramway was converted from single to double track with the exception of the sections in Stodhart tunnel and under Buxton Road.
1803	To cater for the considerable amounts of water that would be required once the Marple Locks were in use, Oldknow together with George Benson Strutt of Belper and Richard Arkwright of Cromford purchased land for reservoir. This land had been earmarked by the Canal Company for several years and a small draft map of 11th November, 1793 shows a reservoir to be built on the site.
	On 8th June, 1803, William Bradbury conveyed part of the Newfield Estate to Oldknow & partners for £1302 12s. 6d.
1804	(13th Oct.) Marple Locks completed and the line was now open for boats from the Ashton Canal to Bugsworth.
1805	Additional land purchased at Combs for construction of parts of the Canal Banks and sluices, area now purchased amounted to 13 acres and with the 1803 purchase made a total of 31 acres.

1809	(9th Nov.) Conveyance of the reservoir & land to the canal company, Arkwright & Oldknow to retain rights of the osier beds, fish and wildfowl.
1810	The Grand Junction Canal Company proposed the construction of a canal across the Peak to link the Cromford and Peak Forest Canals. this scheme did not materialize, but see 1825.
1813	(Sept.] William Chapman (1769-1833), an engineer of canals and railways, proposed that the towns of Sheffield and Manchester be linked by means of a canal and tramway. The route was for either a tramway or canal from Sheffield up the Sheaf Valley to the foot of the ridge at Shawhey Lean, thence by inclined planes to near East Moor. In doing so the line would reach 440 ft in just over 6 miles, then a 2¾ mile tunnel would pass under East Moor to Padley Mill on the Burbage Brook, whence a canal would be constructed to join the Peak Forest Canal at Chapel Milton.
1815	Joel Hawkyard, a surveyor from Ashton-under-Lyne, produced a plan for a proposed tramway from Beard (near Disley] to Bugsworth and a branch from Clough Bar on the Peak Forest Tramway to Ashopton, this route to be via Sparrowpit, Man Tor & Loose Hill.
1824	(24th June] A prospectus was issued for "The Grand Commercial or Scarsdale and High Peak Canal". This scheme to join the Sheffield, Cromford, Chesterfield and Peak Forest Canals was devised by Joseph Hazelhurst, a colliery engineer of Unstone. It was to some 44 miles in length and to run from the Peak Forest Canal at Bugsworth to Hope, then by the Derwent Valley to Grindleford and on to the head of the Barlow Brook. Here it would split, one line going to meet the Sheffield Canal and the other to connect first with the Chesterfield Canal and then the Cromford Canal, joining the latter at Bucklow Hollow.
1825	(2nd May] "The Cromford and High Peak Railway Act" sanctioned. This Act was for "Making and maintaining a Railway or Tramroad from the Cromford Canal, at or near to Cromford, in the parish of Wirksworth, in the County of Derby, to the Peak Forest Canal at or near to Whaley in the County Palatine of Chester." As originally mooted, the scheme was for a canal to link the existing canals so as to provide a direct line between the Midlands and Manchester, however, due to the difficulty of obtaining an ample supply of water on the high limestone plateau, the canal was changed to a railway. The line as built was designed by Josiah Jessop of Butterley Hall and he used the survey for the canal, substituting inclined planes for locks and allowed the line to follow the contours, twisting and winding across the limestone dome of the Peak.
1826	Sanderson of Porto Bello, Sheffield, issued a prospectus entitled "Considerations on the proposed Communications by a Navigable Canal between the Towns of Sheffield and the Peak Forest Canal." The proposal was for a railway from the Sheffield Canal Basin, via the east bank of the River Sheaf to the Porter Valley, then by two incline planes near Nether Green to Soughby, then over the moors to cross the River Derwent near Ding Bank by means of a high level bridge. Continuing on the south side of the Derwent to Gimber Carr, the route traversed a tunnel (½ mile in length] to Edale, up the Edale Valley followed by a second tunnel under Cowburn (1 ½ miles] to the Roych, then to Chapel Milton to join the Peak Forest Tramway, the cost being estimated at £150,000). (It is interesting to note that the proposed tunnels were later constructed, the one from Gimber Carr to Edale by the then Derwent Valley Water Board, in 1961 to

CHRONOLOGY OF THE PEAK FOREST TRAMWAY AND CANAL 73

	carry water from the Edale Valley to the Lady Bower Reservoir in the Derwent Valley. That under Cowburn was built by the Midland Railway Co., in the 1890s as part of the Dore & Chinley Extension Railway; this tunnel when completed was the 9th longest in Britain at 2 miles 182 yds.)
1827	(April] The "Macclesfield Canal Act" sanctioned. (7 Geo. IV c.30½).
1827	The act for the Macclesfield Canal received Royal Ascent,: 7 Geo IV c30 {April 1827). This canal would provide a direct route to the Stoke on Trent area.
1830	The shares of the Peak Forest Canal Co .. were quoted on the London Stock Exchange at £88 and paid a dividend of £3 per share.
1831	(9th Nov.) The Macclesfield Canal opened – it was 26 miles 8 chains in length linking the Peak Forest Canal at Marple with the Trent & Mersey Canal near to Kidsgrove.
1831	The Macclesfield Canal opened 26miles 8 chains in length.
1834/35	Some 4,010 yards of the Tramway re-laid with new rails (& blocks where Needed).
1845	(8th· Oct.) At the Board Meeting held this day, it was decided to discuss terms with the Sheffield, Ashton-under-Lyne and Manchester Railway Company, for the railway to take a lease of the Peak Forest Canal and Tramway.
1846	(1st Jan.) The principal agent of the Peak Forest Canal Co., James Meadows, became Clerk & Secretary of the Sheffield, Ashton-under-Lyne and Manchester Railway Co. (25th March) The tramway and canal leased to the above railway on an annuity of £9325 per annum (i.e. at 5 per cent interest on the issued shares) plus £1856 to cover interest on debts of £41,000.
1847	(1st Jan.) The Sheffield, Ashton-under-Lyne and Manchester Railway Co., became a constituent part of the Manchester, Sheffield and Lincolnshire Railway Co. (MS&LR Co.) (2nd July) The Railway Co. under an Act of Parliament obtained powers to construct a line from Bugsworth to Bradshaw Edge and to sell surplus water from the canal to Manchester, Salford and Stockport for domestic and other purposes at 2d. per 1,000 gallons. Use is still made of these powers by British Waterways, the water now being sold to commercial users near the Canal.
1854	The Railway & Canal Traffic Act of the year, based on the recommendations of a committee chaired by Mr Cardwell, required railway and canal companies to provide facilities for the interchange of traffic and through booking without undue preference. The effect of this Act was to increase traffic on the Peak Forest Tramway as goods could be sent to areas not served by canal, as well as allowing traffic into the area from the North. (31st July) The Stockport, Disley & Whaley Bridge Railway was sanctioned by Act of Parliament, in spite of considerable opposition from the MS&LR Co., the Warrington & Stockport and the Midland Railway companies, aided by the Duke of Devonshire and other landowners. The Midland had asked the MS & LR to survey a line from Whaley to Buxton using in part the route of the Peak Forest Tramway in an attempt to stop the LNWR entering the Peak District. The Act allowed for a junction to be made with the Cromford and High Peak Railway and for other interested railways to make connections between the 6 mile 4 furlong post and Whaley Bridge. (30th Sept.) Work started on the Stockport, Disley & Whaley Bridge Rly.

1857	(28th May) Official opening of the Stockport to Whaley line. (27th July) The LNWR obtained powers for the line to be carried on from Whaley Bridge to Buxton.
1859	Work started on the Buxton extension.
1863	The Stockport, Disley & Whaley Bridge Extension open to Buxton, with a goods yard at Dove Holes and sidings for the lime works of John Bibbington, thus Bibbingtons traffic was lost to the Peak Forest Tramway.
1863	The Midland Railway's line from Derby to Manchester completed, with stations and goods yards at Bugsworth, Chinley, Chapel-en-le-Frith and Peak Forest; in addition sidings were provided for the quarries in Dove Holes Dale. (2nd Aug.) Full control of the Peak Forest Canal Co. passed to the MS&LR Co.
1870	During the 1870,s, considerable quantities of Crist Quarry stone was transported first by Tramway, then Canal to the interchange with Manchester, Sheffield & Lincolnshire Railway for use at Immingham Docks, then under construction.
1897	The MS&LR Co. became part of the Great Central Railway.
1898	A J. Cotton murdered his wife in the cabin of narrow boat 'Annie' whilst moored at Buxworth. She was buried in Chapel Church graveyard. J. Cotton was executed for the murder on 21 st Dec.
1909	(29th Sept.) The last private haulage teams on the tramway were withdrawn, leaving only three teams owned by the company.
1915	The last working kilns at Bugsworth (4 on the north side near Bugsworth New Road, closed.
1923	The canal company transferred the lease of the New Line Quarry to S. Taylor & Frith.
1924	Line closed between Dove Holes and Chapel Milton as all the limestone traffic was being shipped either by rail or by road haulage. This left only a small amount of coal traffic for the mills between Bugsworth and Chapel Milton plus the occasional load of cloth down to the wharf.
1925	The Great Central Railway was now part of the London & North Eastern Railway Co., and in that company's Act for 1925 a clause authorized the closure of the Peak Forest Tramway.
1926	Only very spasmodic traffic on the canal above Marple and all the canal buildings at Bugsworth closed.
1936	Messrs T.W. Ward of Sheffield started to lift the track for scrap.
1943	Considerable lengths of the trackbed sold.
1946	Canal derelict.
1962	Part of the aqueduct at Marple over the River Goyt collapses, the British Waterways Board wanted to demolish the remainder but due to the actions of Marple, Compstall and Bredbury Councils the aqueduct is restored.
1968	Restoration work started on the Bugsworth Basin by the IWPS. Their first attack was on the length from the stop lock near to Bridgemont up to and including the gauging lock near the old canal offices at Bugsworth. One major task was the "Run-off" weir or overflow into the River Goyt.
1970	The IWPS awarded a Bronze Plaque under the European Conservation year, Countryside Awards Scheme for their work to date.
1974	(13th May) After several years work by the Peak Forest Canal Society in restoring the Marple Flights of Locks and extensive clearing and dredging of the canal by members of the Inland Waterways Association and others, the Ashton and

CHRONOLOGY OF THE PEAK FOREST TRAMWAY AND CANAL

	Lower Peak Forest Canals were officially reopened by the Minister of State at the Department of Environment, Mr Denis Howell.
1977	Site granted status as a Scheduled Ancient Monument.
1979	IWPS awarded £600 under the 1978/79 Shell Inland Waterways Restoration Award Scheme.
1980	The Blackbrook Valley becomes the subject of a by-pass enquiry; despite considerable opposition by the local populace and other bodies including the IWPS, the Inspector authorizes the by-pass.
1984	Work starts on the by-pass.
1987	By-pass opened. Landscaping of the area adjoining the canal and parts of the tramway started.
1988	Work on restoring the Basin at Bugsworth proceeding.
1999	The official re-opening of the Buxworth complex by the Mayor of the High Peak local authority. The local paper stated that some 40 boats turned out. 2000 Still problems with leakage. British Waterways carried out a structural survey.
2003	A contract for relining several sections of the complex with concrete was agreed.
2004	February saw a survey of the route of the tramway by Entec UK Limited to assess the feasibility of re-instating the Peak Forest Tramway for use by walkers, cyclists, horse riders and the mobility impaired. The summary report was published in August 2004.
2004	The second re-opening took place over the Easter weekend - it was a Gala event with around 100 boats visiting the complex.
2006	The Peak Forest Tramway Trail is established between Bugsworth Basin and Chapel-en-le-Frith.

Pleasure boats crowd the main arm of the basin, June 2006.   *Author*

The warehouse at Marple, near lock No. 9 seen here before its conversion into offices etc.
Author's collection

Denton Wharf in 1958 with its disused crane well in evidence.   *Author's collection*

Appendix Two

Some employees of the Canal and Tramway and the Private Haulage Teams

1794-1804	Benjamin Outram	Engineer
1794	Thomas Brown	Resident Engineer
1801	Cotton Joderel	Estate Agent
1804	Robt. Preston	Bookkeeper
	John Wronksley	Agent
1808	Jerman Wheatcroft	Ganger in charge of the Chapel Plane
	James Hill	Brakesman at Plane
	James Meadows	Principal Agent (at Piccadilly St, Manchester)
	William Bate	Clerk-in-charge at Bugsworth Wharf
1823-1889	Jack Eyre	Platelayer, later became Ganger
1824-1894	Francis Fletcher	Joined the Company aged 9, he later became the Stone Agent at Chapel in charge of the Limestone Quarries and the Tramway, when retired age 78, the MS & LR Co. awarded him a pension on his full pay of £125 per annum.
1831	James Wood	Engineer
1835	John Britland	Agent at Chapel
	John Walker	Book-keeper
	Wm. Green	Wharfinger at High Lane
	Henry Washington	Bookkeeper
1848	John Sigley	Manager at Whaley Lime Wharf
1849	John Potter	Canal Agent at Bugsworth
1884	John Chappel	Entered service in the Goods Depot, appointed goods agent at Stockport 1st August, 1896, from 1st Dec., 1903 he was the canal and tramway Agent at Bugsworth transferring to Marple on 20th Aug., 1926 as the traffic from Marple to Bugsworth was negligible as there was no stone traffic left.
1890	Adam Jackson	Blacksmith at Bugsworth
	"Owd" Worth	Paymaster, resided at the Canal House, Bugsworth.
1904	Robert Jamison	Blacksmith, Top O' Plane
1905	Edwin Swain Bagshaw	Joiner Top section
	Joe Marchington	Joiner Top section
	William Cartledge	Joiner Top section
	George Lomas	Blacksmith Top section
	Harry Fletcher	Dayman
	James W. Lomas	Dayman
	Robert Joel	Dayman
1901-1911	Jack Eyre	Platelayer
1903-1912	Samuel Eyre	Lengthsman/Ganger
1910-1920	Philip Marchington	Brakesman at Top O' Plane

Pearsons		Provided two-horse teams for hauling coal from Bugsworth to Whitehall Paper Mills and carried paper back on the return journey. Also hauled stone from Crist Quarry for Joel Goddard, Mason of Chinley.
Greens		Two teams of two or three horses provided haulage for Hadfields Mill, and Bridgeholme Mill, together with a considerable amount of sub-contracting for the Canal Co.
Marchingtons		Two and three horse teams for the quarries and general traffic from the Top of the Plane at Chapel-en-le-Frith to Dove Holes.

Appendix Three

Lime Kilns and Lime Burners on the route of the Canal or Tramway

Marple	1797	Samuel Oldknow built 2 kilns at a cost of £1527 a further 4 kilns added at a cost of £3000
	1835	James Clayton & Co worked the kilns
	1890	J. & M. Tymn
	1891	Buxton Lime Firms
Disley	1810	Farey records lime kilns in use
Bugsworth	1807	Wright & Brown – south side of Brook
	1857	Thomas Boothman & Co.
	1857	Robert Sattersfield – north side of Brook
	1870	Kilns rebuilt
	1890	William Pitt Dixon
	1891	Buxton Lime Firms
Chapel Milton	1830	John & Thomas Brocklehurst
	1860	MS&LR Co.
Dove Holes Dale	1742	Hallsteads owned by the Gisborns, kilns from 1800.
	1808	Knat Hole Kilns, George Potts.
	1877	Joseph Wainwright opened a quarry and erected 8 kilns before 1885, became part of Buxton Lime Firms 1891.
	1880	Perseverance, became part of Buxton Lime Firms closed by 1930.
	1864	Joel Carrington a lime merchant of Hollinwood, built 2 kilns, when he died in 1878 his executors sold them as a going concern to Samuel Taylor; Holderness Quarry opened and a further 8 kilns erected.

It is interesting to note that in July 1650 a General Survey of the Manor of the High Peak mentions limestone and lime burning in Dove Holes "All those Quarries or Pits of lymestone lying in ye Crofts by ye Dovehole near Chapell Frith within the wast grounds of the Manor aforesaid for the burning whereof there are at present 14 Kilns at work."

	1895	Bold Venture Lime Works: Gaskell, Deacon & Co. Small Dale Lime Works: Thos. Beswick & Son.

Appendix Four

Other Lines that served the Canal

Under a clause in the original Act for the canal:

> Proprietors of mines of coal, stone, furnace or other works are empowered to make branch railways of not more than 2000 yards in length and branch cuts or canals of not more than 4 miles in length to communicate with the canal.

Use was made of this clause to construct several lines to the canal, but the first was built under the general powers of the Act:

1. THE MARPLE TRAMWAY

    A tramway was constructed as a temporary link to join the two levels of the canal whilst funds were being raised to enable the locks to be built. According to the Peak Forest Canal Minute Book the Marple Tramway was in use by 1800 and continued so until the locks were completed and was in fully operational order in 1807. The length was 1½ miles; it was built as an inclined railway, loaded wagons descending, hauling up empties. It was similar to the Peak Forest Tramway, using L-type plates and trucks with detachable bodies.

2. THE MARPLE LIME KILNS RAILWAY

    A railway was built from lock 12 to the lime kilns built by Samuel Oldknow. Farey states:

    > Where 4 locks occur near together on the canal a branch has been taken out of the upper pound to a dock, where stone and coal boats lie to unload. 12 kilns can burn 2500 bushels of lime daily.
    >
    > From the bottom of the lime kilns, railways are laid and constructed, some into a boathouse over two boats that can lie in a dock connecting with a lower pound of the canal and having their lading of lime tippled or turned over into them from the trams on the railway, under cover of rain, others of the railways are conducted into a lime-house over four or five carts or wagons that can stand at the same time and have the lime tippled into them, secure from the weather, and others to tipplers without cover.
    >
    > The Lime Kilns are rather egg-shaped, 36 feet deep, 13½ feet diam. at the top and 14½ feet in the belly or widest place at 9 feet down, diminishing thence to 3½ feet diam. at the bottom. Iron sheets are used to draw the lime at 20 inches above the floors on which the railways are laid. Between the bottoms of the kilns, roomy arched stables are constructed, in some of which the farmers feed and rest their horses, whilst their carts are being loaded and others are let to boat-men for their towing horses.
    >
    > Mr. Oldknow purchases limestone at Bugsworth Wharf at 2s. 1d. per ton of 2400 pounds and coals are delivered at the kilns at 5s. to 6s. 8d. per ton. The lime is sold at 14d. per load of ten winchester level pecks.

    The lime produced at Marple was sold 50 per cent locally for agriculture, cement, bleaching and chemicals, and the other 50 per cent was shipped by the canal to Bolton, Bury, Rochdale and Saddleworth.

The kilns and railways were later taken over by J. & M. Tymn, who in turn in 1891 sold out to the Buxton Lime Firms Co. Ltd. A plan of 1891 shows a railway from the top pound to the mouth of the kilns, as well as railways from the bottom to the canal.

3. DISLEY KILN TRAMWAY
   A short line from the canal wharf to the top of the kilns, opened by 1810, most likely used detachable bodied wagons as used on the Peak Forest Tramway.
4. DIGLEE RAILWAY
   A tramway from Jowhole to pits at Diglee, approx. 1 mile in length.
5. YEARSLEY TRAMWAY
   A tramway from a wharf under the road to stone quarries at Yearsley. The tunnel under the road is still in existence.
6. BUGSWORTH
   Lines were laid at various periods from the many small coal pits on the hillsides of the Lyme Valley.
6A. A line from the Dolly Pit to William Pitt Dixon's Lime Kilns at Bugsworth where there would have been a connection with the Peak Forest Tramway. The route is shown on a copy of the plan deposited by the Midland Railway in 1860. Route still visible in parts.
6B. A line from Higher Portobello Pits to Barren Clough and possible connection with Peak Forest Tramway. Shown on 6 in. OS leading to a land sale wharf.
6C. A line from Mosley Hall Pits to Basin pre 1850.
6D. A line from Bugsworth Pit to Basin prior to 1860.

The aqueduct and viaduct over the River Goyt near Marple. The photograph was taken in 1962 shortly after repairs to the aqueduct and with the canal back in water.

Author's collection